Anthropogenic Solutions for Climate Change

Olivier Serrat

Anthropogenic Solutions for Climate Change

Achieving Environmental Peace

Springer

Olivier Serrat
School of Continuing Studies
Georgetown University
Washington, DC, USA

ISBN 978-981-96-8653-7 ISBN 978-981-96-8654-4 (eBook)
https://doi.org/10.1007/978-981-96-8654-4

This Springer imprint is published by the registered company Springer Nature Singapore Pte Ltd.
The registered company address is: 152 Beach Road, #21-01/04 Gateway East, Singapore 189721, Singapore

This book is dedicated to those who remain steadfast in their commitment to ensure a sustainable future for generations to come. Each contributes in their own way, yet all are united by their shared purpose.

To marginalized and vulnerable communities, who endure the harshest impacts of climate change yet inspire hope through their resilience and strength.

To the scientists, researchers, policymakers, educators, and communicators, whose efforts deepen understanding, raise awareness, and drive actionable strategies, including the preservation of biodiversity as a cornerstone of sustainability.

To activists, especially young leaders, who challenge norms and push for meaningful progress.

To visionaries, who turn ambitious ideas into practical solutions, advancing fairness for all.

To collaborators, who take bold action for environmental peace, overcoming self-interest.

And to everyone who stands firm in the pursuit of environmental justice, building a flourishing and harmonious planet.

May each reader of this book see themselves reflected in these words and feel inspired to continue their essential work.

Anthropogenic Solutions for Climate Change: Foreword

Environmental challenges, including climate change and air pollution, deforestation, water crises, land degradation, waste management, and biodiversity loss, pose significant threats to civilization and require urgent action to protect ecosystems and sustain life on Earth. These challenges have intensified with economic growth and population increases. Environmental problems affect both rural and urban areas, particularly in the rapidly growing megacities and water-scarce regions undergoing desertification in developing countries. However, North America is not immune to environmental catastrophes, as the Dust Bowl during the Great Depression and the Love Canal disaster demonstrate. Moreover, global warming and greenhouse gases pose a threat to everyone on the planet. There is a strong nexus between environmental degradation, poverty, and severe income inequality. Unfortunately, existential environmental threats are not universally acknowledged, including by some in powerful positions in wealthy countries.

Like most individuals from the post-World War II baby boomer generation, my understanding of the need for a strategic approach to environmental issues developed gradually. It was first stimulated by reading Jared Diamond's *Guns, Germs, and Steel: The Fates of Human Societies* (Diamond 1997) and *Collapse: How Societies Choose to Fail or Succeed* (Diamond 2005). These books demonstrated that some societies collapsed due to environmental factors. The latter presented thought-provoking questions such as why some societies make catastrophic decisions that lead to their downfall, how businesses affect the environment, and what individuals can do to help address environmental problems.

My understanding of environmental issues deepened as my professional career progressed, particularly during my 11-year tenure managing the Asian Development Bank's portfolio in China. At the time, China faced many severe environmental challenges. With an annual lending program of about $1 billion, we tackled a broad range of environmental issues—air pollution, water pollution, land degradation, energy conservation, renewable energy, environmental policy, and regulation—from the 1990s through 2004. That experience taught me that environmental issues can be successfully addressed through investment in clean technology and infrastructure, promotion of a circular economy, implementation of appropriate price signals, and improved regulation. Based on this practical experience, I firmly believe in "win–win" solutions to environmental issues—advancing environmental protection while fostering economic growth, higher incomes, and a better quality of life.

During my tenure from 2005 to 2008 as Director General of the Asian Development Bank's Operations Evaluation Department, I learned the critical role of monitoring and evaluation in identifying what works and what does not. Systematic, data-driven analysis is essential for shaping practical, effective climate policies. Lessons learned from development projects over decades underscore that success in combating climate change depends on rigorous monitoring, accountability, and iterative refinement of strategies. A commitment to feedback and evaluation is essential for developing and implementing impactful, scalable solutions.

During that period, I worked directly with Olivier Serrat, whom I sought out for his hands-on expertise in addressing the nexus between the environment and poverty through projects across many countries. Notably, he tackled environmental sustainability and livelihood issues in the Tonle Sap region, Southeast Asia's largest freshwater lake and a vital part of the Mekong Basin. At the time, a strategic approach to knowledge management was in its infancy at the Asian Development Bank. Olivier played a pioneering role in developing a strategy that integrated knowledge management and evaluation results to improve future operations. This work culminated in the publication of *Learning in Development* (Serrat 2010). Olivier later became head of the Asian Development Bank's Knowledge Management Center, where he advanced the use of knowledge management to optimize operations and development impact.

Anthropogenic Solutions for Climate Change is a compelling call to action, backed by thorough research and sound analysis. The world must take timely and decisive action to address climate change—it is not a distant threat. Climate change is real and a direct consequence of human activity. While humanity has accelerated environmental degradation and its march toward the end of civilization as we know it, we can reverse this trajectory through deliberate, innovative policymaking and development strategies. However, our window for effective intervention is narrowing.

The urgency of action must drive decisions across governments, financiers, consumers, and society at large. We must envision a future where economic progress and environmental protection coexist, fostering sustainable development. Achieving this will require shifting away from traditional development paradigms—moving from extractive practices to regenerative systems that promote prosperity while restoring ecological balance.

This book is a genuine tour de force. Its strategic themes, conclusions, and recommendations are grounded in a wealth of real-world experience and detailed case studies. The section on Climate Change and Development, for example, analyzes Indonesia's climate change mitigation strategy and Vietnam's climate adaption strategy. The section on Climate Change and International Security presents a compelling case that China and the United States must cooperate for the world to successfully address climate change. It underscores that climate security is deeply linked to US national and international security. These findings support the call for a modern-day Marshall Plan for climate adaptation and climate justice. Other critical strategic issues explored in this book include building sustainable futures, deglobalization and public diplomacy, deforestation, environmental migration, environmental peacebuilding, localizing climate action in Latin American and Caribbean cities, and the environment in Africa.

Human ingenuity, creativity, innovation, and determination know no bounds. Achieving transformative innovation at scale is critical to successfully addressing global warming and achieving environmental peace, safeguarding civilization. Olivier is uniquely qualified to write this landmark work. His career at the Asian Development Bank centered on practical, real-world issues spanning the environment, poverty, and sustainable livelihoods in developing countries, as well as strategy development and implementation, evaluation, and knowledge management. Following that, he spent several years as an academic, researcher, and speaker specializing in two key fields—organizational leadership and environmental management. He has recently published four distinguished books: *Leading Organizations of the Future: A New Framework* (Serrat 2023a), *Digital Solutions: Reframing Leadership* (Serrat 2023b), *Leading Solutions: Essays in Business Psychology* (Serrat 2021), and *Knowledge Solutions: Tools, Methodologies, and Approaches to Drive Organizational Performance* (Serrat 2017). Further strengthening his expertise, he earned a late-in-life PhD in organizational leadership from the Chicago School of Professional Psychology, demonstrating his lifelong commitment to learning. Olivier combines practical experience with academic rigor, equipping him to develop a comprehensive, effective approach to tackling the greatest challenge facing civilization.

Following Olivier's advice will help civilization avoid the fate forewarned by Paul Bahn and John Flenley: "Humankind's covetousness is boundless. Its selfishness appears to be genetically inborn [...] But in a limited ecosystem, selfishness leads to increasing population imbalance, population crash, and ultimately extinction." (Bahn and Flenley 1992).

Vancouver, BC, Canada

Bruce Murray

References

Bahn P, Flenley J (1992) Easter Island, Earth Island: the enigmas of Rapa Nui. Thames & Hudson.
Diamond J (1997) Guns, germs, and steel: the fates of human societies. W. W. Norton.
Diamond J (2005) Collapse: how societies choose to fail or succeed. Viking Press.
Serrat O (2010) Learning in development. Asian Development Bank.
Serrat O (2017) Knowledge solutions: tools, methods, and approaches to drive organizational performance. Springer.
Serrat O (2021) Leading solutions: essays in business psychology. Springer.
Serrat O (2023a) Leading organizations of the future: a new framework. Springer.
Serrat O (2023b) Digital solutions: reframing leadership. Palgrave Macmillan.

Preface

"Everybody talks about the weather, but nobody does anything about it," Mark Twain reportedly quipped. In the era of climate change, it is time to flip that script. While Mark Twain's remark was a witty observation on inaction, a paraphrase of Leon Trotsky elevates the urgency of the matter: "You may not be interested in climate change, but climate change is interested in you." This reimagined observation shifts the narrative entirely. Today, we must go beyond discussing the weather—a metaphor for climate change—but also commit to decisive action to address its pressing challenges. Serrat (2012) documented global sectoral and thematic milestones toward sustainable development, tracing progress as far back as 1948. However, as the pressures of climate change escalate, accelerating efforts in both climate change adaptation and mitigation have become imperative. While some effects of climate change are irreversible (e.g., rising temperatures, melting ice caps and glaciers, sea level rise, extreme weather events, ocean acidification, loss of biodiversity, agricultural impacts, and health risks), the steps we take today and tomorrow still hold the power to make a significant difference. The stakes could not be higher—the future of our planet hangs in the balance.

This brings us to a crucial question: Can humanity rise to the challenge of climate change? *Anthropogenic Solutions for Climate Change: Achieving Environmental Peace* explores this existential issue. The power to make a difference lies within each of us. This book celebrates human ingenuity and collaboration, showcasing innovative strategies to navigate the critical challenges of climate change. It also examines past failures and shortcomings, providing valuable lessons to guide and shape future initiatives.

Given the increasing frequency and intensity of climate-related disasters, it has become clear that immediate and sustained efforts are essential. This book bridges the gap between climate science and implementable policy, offering professionals and practitioners the tools they need to mitigate or adapt to the impacts of climate change. While development, international security, public diplomacy, environmental migration, environmental peacebuilding, and ecotopian visions are not the only tools in the fight against climate change, they remain crucial. A comprehensive, multidimensional approach—integrating diverse fields—is required to effectively address this global challenge.

As you explore this book, consider the interconnectedness of our world and the shared responsibility we hold in tackling climate change. Each chapter is designed to provide insights and practical solutions for various contexts. The modular structure allows chapters to be read in any order, offering enhanced accessibility, ease of reference, engagement, and reusability. Whether you are a policymaker, researcher, or concerned citizen, this book seeks to inspire action on all fronts, empowering individuals and institutions to contribute meaningfully to a sustainable and peaceful future. Let us update Serrat (2012) for the better.

Washington, DC, USA Olivier Serrat

Reference

Serrat O (2012) World sustainable development timeline. Asian Development Bank.

Competing Interests The author has no competing interests to declare that are relevant to the content of this manuscript.

Contents

List of Figures

List of Tables

Part I
Climate Change & Development

Chapter 1
Citius, *Altius*, *Fortius*: A Climate Change Mitigation Strategy for Indonesia

Abstract As one of the top ten largest emitters of greenhouse gases, Indonesia's intensified actions are crucial for significantly impacting global climate goals and protecting its vulnerable ecosystems and communities. This précis details an urgent climate change mitigation strategy for Indonesia, emphasizing the need for immediate and sustained efforts to ensure both local and global climate stability.

1.1 Indonesia's Climate Mitigation Imperative

As accelerating climate change impacts the world, Indonesia stands at a pivotal moment. Intensifying mitigation efforts is not just a global and moral responsibility; it is a national imperative to safeguard the future and well-being of every Indonesian (Serrat 2024a).

Although developed nations have historically been the largest contributors to greenhouse gas emissions, developing countries must also engage in climate change mitigation. This global responsibility is crucial to protect those most at risk, such as low-income families, marginalized communities, indigenous peoples, and coastal residents. Climate justice demands support for populations who have contributed the least to climate change but suffer the most. By participating in mitigation, Indonesia can uphold climate justice, ensuring an equitable transition to a low-carbon economy while safeguarding vulnerable communities.

The Paris Agreement aims to unite countries in combating climate change by limiting global warming to well below 2 °C above pre-industrial levels, with efforts to limit the increase to 1.5 °C (United Nations 2015). Currently, global average temperatures are about 1.3 °C higher. The world has not surpassed the Paris Agreement's 1.5 °C goal as it is based on long-term average temperatures, but the first nine months of 2024 are likely to have breached the threshold. As the world's 16th largest economy, Indonesia ought to take an active role in achieving the Paris Agreement's goal: it is pointless to remain inactive and blame others while one's own home is ablaze. However, the Climate Action Tracker (2024) ranks Indonesia's performance as "Critically Insufficient." Despite pledging to phase out coal at

O. Serrat, *Anthropogenic Solutions for Climate Change*,
https://doi.org/10.1007/978-981-96-8654-4_1

COP26 in 2021, Indonesia increased coal consumption by 33% in 2022, leading to a 20% rise in emissions and making it the world's sixth-highest emitter (Jong 2023). Deforestation remains a major contributor to Indonesia's emissions, with the land-use and energy sectors accounting for 84% of the country's total emissions (USAID 2022).

International agreements and funding mechanisms provide financial and technical assistance for a sustainable future. By engaging in mitigation, Indonesia would signal global cooperation, boost its international standing, attract foreign investments, and benefit from global support and technology transfers. Promoting international cooperation and solidarity would ensure a more equitable and sustainable distribution of resources and opportunities for Indonesia's population.

Notwithstanding local pollution issues like transboundary haze, water pollution in the Citarum River, urban air pollution, and plastic waste, Indonesia's population remains highly vulnerable to climate change. This vulnerability is highlighted by the government's decision to relocate its sinking capital to Nusantara in eastern Borneo (Serrat 2024b, c). Indonesia ranks fifth globally for the number of people living in low-lying coastal areas (World Bank & Asian Development Bank 2021). The country faces frequent and severe weather events that damage agricultural lands, homes, and infrastructure. Coastal erosion and saltwater intrusion displace communities, while shifting rainfall patterns affect agricultural productivity and water availability. Health issues have surged due to more frequent heat waves and vector-borne diseases like dengue fever and malaria. Additionally, warmer waters are bleaching coral and impacting fish populations.

The Islamic Declaration on Global Climate Change (2015) is based on Islamic principles advocating for a balanced, sustainable existence, emphasizing justice, moderation, and respect for all forms of life. By integrating these values into mitigation efforts, Indonesia's substantial Muslim population can substantially contribute to global sustainability and environmental preservation.

1.2 Strategic Actions for Climate Change Mitigation in Indonesia

Indonesia's climate challenges are closely linked to its growth and development. The government's Long-Term Strategy for Low Carbon and Climate Resilience aims to balance environmental sustainability with economic growth (United Nations Climate Change 2021; World Bank 2023). However, Indonesia's Enhanced Nationally Determined Contribution targets only a 32% reduction in emissions below the business-as-usual scenario by 2030 unconditionally, and a 43% reduction conditionally (United Nations Climate Change 2022). The government has set a distant net zero target for 2060, but its comprehensiveness remains unrated due to incomplete information. Emissions in sectors other than forestry are projected to triple by 2030 (Climate Action Tracker 2024). In short, Indonesia's ambitions are insufficient to meet the Paris Agreement goals (Climate Action Tracker 2024; United Nations 2015).

To build a resilient nation that thrives in harmony with nature and ensure a better life for all Indonesians, including future generations, Indonesia must adopt a bold mitigation strategy, moving faster, aiming higher, and growing stronger. The strategic thrusts outlined below will appeal to domestic audiences, climate justice advocates, and the international community. The following specific actions should be prioritized:

- Transition to Renewable Energy
- Forest Conservation and Restoration
- Public Transportation Enhancement
- Carbon Pricing
- Energy Subsidy Reform

To clarify, some of the strategic thrusts outlined below that have to do with sustainable agriculture and water resource management—specifically adaptive farming practices and agri-food system integration as well as sustainable water management—are primarily linked to climate change adaptation. These strategic thrusts are included in the climate change mitigation strategy to enhance the political acceptability of the overall package.

Key roles are expected from the Ministry of National Development Planning (BAPPENAS), the Fiscal Policy Agency in the Ministry of Finance, the Ministry of Energy and Mineral Resources, the Ministry of Environment and Forestry, and city councils. New governance structures, such as Chile's Climate Change Framework Law, could decentralize responsibility by assigning climate-related tasks to multiple government ministries, regional authorities, and municipal governments. Inclusive approaches like South Africa's Just Transition Framework would promote a shared vision, principles, policies, and governance arrangements to achieve net zero emissions while ensuring that vulnerable communities are not harmed in the process. Learning by doing is a powerful approach to climate action. Figure 1.1 illustrates the key components of a climate change mitigation strategy for Indonesia.

1.2.1 Environmental Sustainability

1.2.1.1 Transition to Renewable Energy

- **Coal Phase-Out**. Phase out coal usage and invest in renewable energy.
- **Renewable Energy Focus.** Emphasize solar, wind, and geothermal energy to diversify the energy mix and reduce emissions.

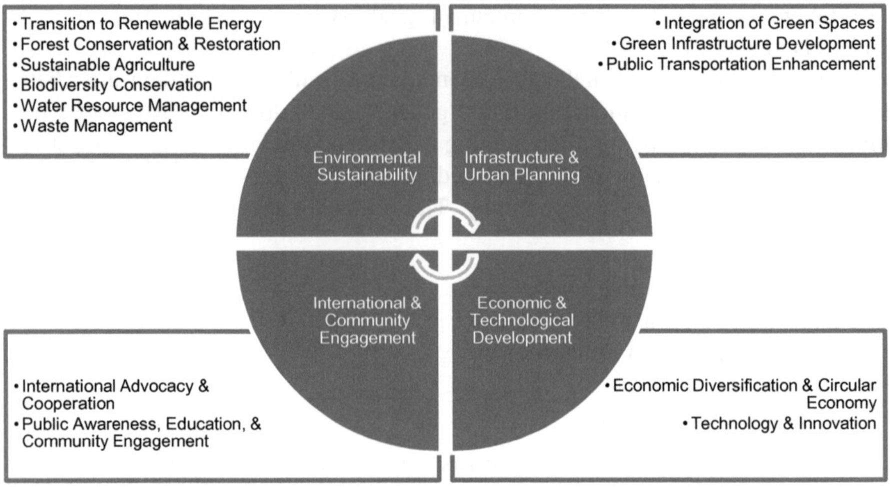

Fig. 1.1 A climate change mitigation strategy for Indonesia

1.2.1.2 Forest Conservation and Restoration

- **Deforestation Prevention.** Strengthen policies to prevent illegal logging and promote reforestation. Aim for net zero emissions in forestry by 2030.
- **Peatland Restoration.** Restore degraded peatlands to prevent emissions and enhance biodiversity.

1.2.1.3 Sustainable Agriculture

- **Adaptive Farming Practices and Agri-Food System Integration.** Promote sustainable farming and integrate agri-food systems in mitigation plans to increase resilience to climate impacts.
- **Agroforestry Integration.** Integrate trees into agriculture to enhance carbon sequestration and soil health.

1.2.1.4 Biodiversity Conservation

- **Protected Areas Expansion.** Expand and manage protected areas to conserve biodiversity.
- **Species Protection Measures.** Implement measures to protect endangered species and habitats.

1.2.1.5 Water Resource Management

- **Sustainable Water Management.** Implement practices to ensure water security and reduce climate impacts.

1.2.1.6 Waste Management

- **Recycling Programs.** Increase efficiency and reach of recycling initiatives.
- **Waste-to-Energy Technologies.** Utilize waste for energy generation to reduce emissions.

1.2.2 Infrastructure and Urban Planning

- **Integration of Green Spaces.** Reduce the urban heat island effect to decrease the demand for air conditioning, leading to lower energy use and emissions.
- **Green Infrastructure Development.** Develop urban forests, green roofs, and parks to enhance carbon sequestration. Green roofs and walls can provide insulation, reducing the need for cooling in buildings, which in turn lowers energy consumption and emissions.
- **Public Transportation Enhancement.** Enhance public transportation to reduce emissions.

1.2.3 Economic and Technological Development

1.2.3.1 Economic Diversification and Circular Economy

- **Promotion of Green Jobs.** Create green jobs in renewable energy and sustainable sectors.
- **Resource Efficiency.** Promote a circular economy to minimize waste.

1.2.3.2 Technology and Innovation

- **Investment in R&D.** Invest in low-carbon technologies and climate-resilient practices.
- **Utilization of Digital Solutions.** Use digital technologies for efficient environmental management.

1.2.4 International Cooperation and Community Engagement

1.2.4.1 International Advocacy and Cooperation

- **Leadership in Climate Forums.** Advocate for stronger global climate action.
- **Global Partnerships.** Share knowledge, technology, and resources for climate action.
- **Regional Cooperation.** Collaborate on regional climate initiatives.

1.2.4.2 Public Awareness, Education, and Community Engagement

- **Nationwide Public Awareness Campaigns.** Educate the public about climate change.
- **Climate Education in Schools.** Integrate climate change topics into school curriculums.
- **Local Community Involvement.** Engage local communities in climate action plans.

1.3 Policy Recommendations for the Government to Drive Economic Action on the Climate Change Mitigation Strategy

1.3.1 Financial Mechanisms and Incentives

- **Funding Mechanisms.** Promote green bonds and climate funds.
- **Carbon Pricing.** Introduce mechanisms to incentivize low-carbon technologies.
- **Energy Subsidy Reform.** Provide subsidies for renewable energy to reduce emissions.

1.3.2 Private Sector Engagement

- **Investments and Incentives.** Encourage private sector investment in green technologies.
- **Public–Private Partnerships.** Foster collaborations for mitigation initiatives.

1.3.3 Regulatory Frameworks

- **Emissions Regulations.** Implement strict emissions regulations.
- **Energy Efficiency Standards.** Set targets for energy efficiency and renewable energy adoption.
- **Waste Management Policies.** Develop comprehensive waste management policies and associated regulations.

1.3.4 Research and Development

- **Investment in R&D.** Advance green technologies and innovative solutions.

1.4 Cost Projections: Handling Unknowns

Indonesia's Enhanced Nationally Determined Contribution projects the country will need at least $310 billion to reduce emissions by 2030. The government has allocated about 4.1% of its state budget to emissions reduction, indicating the need for additional support from corporations, individuals, and other stakeholders. However, according to the World Bank (2021), climate change could cost Indonesia 2.5–7% of its gross domestic product, with the poorest bearing the brunt. Clearly, the current budget allocation falls short of addressing the potential costs of climate change, and the price of inaction far outweighs the proposed expenditures. By comparing costs and budgets with the mitigation efforts of nearby countries, Indonesia can gain insights about its bottom line. As the steward of Indonesia's well-being, the government must safeguard its citizens from the threats of climate change. The urgency of this responsibility cannot be overstated, as the impacts of a warming planet are already being felt across communities, the economy, and Indonesia's ecosystems. Building a new climate economy presents a once-in-a-lifetime opportunity that Indonesia should prioritize.

References

Climate Action Tracker (2024) Indonesia. https://climateactiontracker.org/countries/Indonesia/targets/

Islamic Foundation for Ecology and Environmental Sciences (2015) Islamic declaration on global climate change. http://islamicclimatedeclaration.org/islamic-declaration-on-global-climate-change/

Jong H (2023) Indonesia's coal burning hits record high—and 'green' nickel is largely why. Mongabay, July 3. https://news.mongabay.com/2023/07/Indonesias-coal-burning-hits-record-high-and-green-nickel-is-largely-why/

Serrat O (2024a) ICYMI: climate security rhymes with national security. Unpublished manuscript, Georgetown University. https://www.researchgate.net/publication/378545941_ICYMI_Climate_Security_Rhymes_with_National_Security

Serrat O (2024b) Environmental migration in the Anthropocene: perspectives on the relocation of Indonesia's capital city to Eastern Borneo. Unpublished manuscript, Georgetown University. https://www.researchgate.net/publication/380537926_Environmental_Migration_in_the_Anthropocene_Perspectives_on_the_Relocation_of_Indonesia's_Capital_City_to_Eastern_Borneo

Serrat O (2024c) Preserving paradise: the heart of Borneo Initiative. Unpublished manuscript, Georgetown University. https://www.researchgate.net/publication/381993897_Preserving_Paradise_The_Heart_of_Borneo_Initiative

United Nations (2015) Paris agreement. https://unfccc.int/sites/default/files/english_paris_agreement.pdf

United Nations Climate Change (2021) Indonesia Long-term strategy for low carbon and climate resilience 2050. https://unfccc.int/sites/default/files/resource/Indonesia_LTS-LCCR_2021.pdf

United Nations Climate Change (2022) Enhanced nationally determined contribution: Republic of Indonesia. https://unfccc.int/sites/default/files/NDC/2022-09/23.09.2022_Enhanced%20NDC%20Indonesia.pdf

USAID (2022) Indonesia climate change sheet. https://www.usaid.gov/sites/default/files/2023-03/2022-USAID-Indonesia-Climate-Change-Country-Profile.pdf

World Bank (2021) Climate change knowledge portal: Indonesia. https://climateknowledgeportal.worldbank.org/country/Indonesia

World Bank (2023) Indonesia—Country climate and development report. Washington, D.C. https://www.worldbank.org/en/country/Indonesia/publication/Indonesia-country-climate-and-development-report

World Bank & Asian Development Bank (2021) Climate risk profile: Indonesia. https://www.adb.org/publications/climate-risk-country-profile-Indonesia

Chapter 2
To Baku and Beyond: A Climate Change Adaptation Strategy for Vietnam

Abstract In light of escalating climate extremes, this précis outlines a proposed climate change adaptation strategy for Vietnam. The proposed strategy establishes guiding principles, lists priorities, addresses uncertainties, provides rough cost estimates for specific tasks and solutions, and explores financing options. By highlighting Vietnam's climate vulnerability, the strategy would strengthen the country's case for accessing the Loss and Damage Fund established at COP27.

2.1 COP29: Why It Matters[1]

The 29th Conference of the Parties (COP29) to the United Nations Framework Convention on Climate Change (UNFCCC) will be held in Baku, Azerbaijan, from November 11 to 22, 2024 (UNFCCC 2024). COP29 will build on the findings of the inaugural Global Stocktake concluded at COP28 in 2023, which underscored both progress and significant gaps in global climate action. In Baku, Azerbaijan, countries are expected to set more ambitious targets for reducing greenhouse gas emissions and update their Nationally Determined Contributions, enhance climate change adaptation measures, and boost finance, technology, and capacity building in developing countries that are particularly vulnerable to the adverse effects of climate change (Serrat 2024). In light of escalating climate extremes, this précis outlines a proposed climate change adaptation strategy for Vietnam. The proposed strategy establishes guiding principles, lists priorities, addresses uncertainties, provides rough cost estimates for specific tasks and solutions, and explores financing options. By highlighting Vietnam's climate vulnerability, the strategy would strengthen the country's case for accessing the Loss and Damage Fund established at COP27.

[1] *To Baku and Beyond: A Climate Change Adaptation Strategy for Vietnam* was completed on October 17, 2024 ahead of COP29.

O. Serrat, *Anthropogenic Solutions for Climate Change*,
https://doi.org/10.1007/978-981-96-8654-4_2

2.2 Vietnam's Escalating Climate Crisis: Current Struggles and Future Threats

Vietnam faces worsening climate extremes, as evidenced by Typhoon Yagi in September 2024, which claimed at least 152 lives, injured over 700 people, and caused an estimated $1.6 billion in damages. Typhoon Yagi was the strongest storm to hit Vietnam in 30 years. Vietnam's vulnerability is exacerbated by its extensive coastline, dense population, and dependence on agriculture:

- **Extensive Coastline.** Vietnam's 3260-kilometer coastline and over 3000 islands are highly exposed to rising sea levels, coastal erosion, and salinity intrusion.
- **Dense Population.** The Red River Delta and Mekong River Delta have high population densities, increasing the risk of climate impacts.
- **Dependence on Agriculture.** The low-lying Mekong River Delta is crucial for rice production and home to 18 million people.

In 2020, Vietnam incurred losses of approximately $10 billion due to climate change impacts (World Bank 2022). Without adaptation measures, these losses could mount, potentially pushing up to one million people into extreme poverty by 2030 and costing Vietnam 12–14.5% of its gross domestic product (GDP) annually by 2050 (World Bank 2022).

2.3 A Climate Change Adaptation Strategy for Vietnam

The scientific evidence of climate change is unequivocal, making preparation essential. However, the exact timing and severity of climate change impacts remain uncertain. This uncertainty, coupled with limited financial and other resources, demands that Vietnam should make trade-offs. This précis proposes a flexible strategy that aligns with and integrates into existing national policies and development plans to enable Vietnam to effectively respond to uncertainty.

2.3.1 Principles

The proposed strategy underscores the importance of rapid, robust, and inclusive development. It is guided by principles designed to (i) facilitate the adaptation of firms and individuals; (ii) adapt land use plans and safeguard critical public assets and services; (iii) assist firms and individuals in managing residual risks and natural disasters; and (iv) address financial and macrofiscal risks (Hallegatte et al. 2020). Additionally, the strategy emphasizes locally led adaptation to ensure equitable access to financing and decision-making power for frontline actors (World Resources Institute 2022). The effective application of the strategy hinges on proactive prioritization, diligent implementation, and continuous progress monitoring. It is crucial to maintain a balance between strategic planning and adaptive learning throughout the process (Serrat 2017a).

2.3.2 *Strategic Thrusts*

Climate change adaptation builds resilience in natural, social, and economic systems. In Vietnam, agriculture, biodiversity, physical infrastructure, public health and safety, residential areas, and water resources are particularly susceptible to natural disasters and extreme weather events. The proposed strategy's primary goal would be to mitigate climate change impacts across Vietnam's mainland and territorial waters. Given Vietnam's unique geographic and socio-economic characteristics, the strategy would prioritize a no-regrets approach focused on:

- Climate monitoring and early warning of natural disasters
- Natural disaster response, flood prevention for major cities, and reinforcement of river dikes, sea dikes, and reservoir safety
- Food security
- Water security
- Climate-resilient communities
- Protection and sustainable development of forests and biodiversity

Where the no-regrets approach is not applicable, uncertainty would be treated with reversible policies, safety margin strategies, and soft strategies including insurance and rolling plans (Hallegatte 2009). Wherever feasible, nature-based solutions—such as reducing the impacts of storms through mangrove restoration and decreasing air pollution by establishing green spaces in cities—would be leveraged to promote human well-being and protect biodiversity.

2.4 Implementing Vietnam's Climate Change Adaptation Strategy

2.4.1 *Specific Tasks and Solutions*

Aligning with the United Nations' Sustainable Development Goals, Vietnam's National Adaptation Plan (NAP) for 2021–2030 outlines specific tasks and solutions that can help implement the proposed strategy (Ministry of Natural Resources and the Environment 2022; United Nations 2015). The Ministry of Natural Resources and Environment (MONRE) would lead and coordinate relevant ministries, sectors, and the People's Committees of provinces and centrally run cities to ensure the strategy's implementation. In 2030, MONRE would review and assess the strategy for relevance, coherence, effectiveness, efficiency, impact, and sustainability (Organisation for Economic Co-operation and Development 2024).

2.4.2 *Cost Estimates*

Based on the NAP, the total cost of implementing the proposed strategy is projected to be 3–5% of GDP by 2030. For 2021–2030, an estimated $54.99–$91.65 billion would be required, based on the 2020 net present value and a discount rate of 10%. (There may be a need to triangulate and validate the NAP's estimate with the costs of actual interventions in other countries.) Despite prioritizing the most vulnerable people, regions, and sectors, the financial resources required would exceed what Vietnam's economy can afford. Vietnam should plan to allocate up to 1.5% of its GDP annually to climate change adaptation, totaling $27.5 billion. However, an additional $2.75–$6.42 billion per year from non-state sources would be called for, amounting to $27.5–$64.16 billion over the decade. Toward this, Vietnam would develop mechanisms and incentives to encourage the engagement of businesses and individuals in investment, research, and implementation of climate change adaptation activities. To foster collective responsibility towards climate resilience, Vietnam would also look to engage individuals through public awareness campaigns, community involvement in planning and implementing adaptation projects, financial incentives for sustainable practices, volunteer programs, feedback mechanisms for citizen input on climate policies, and comprehensive educational programs.

2.4.3 *The Need for International Support*

Other countries can support Vietnam's climate change adaptation efforts in several ways: capacity building, financial aid, policy support, partnerships, and technical assistance (Serrat 2017b, c). The priority areas for international support would be (i) strengthening policy capacity and human resources for climate change adaptation; and (ii) enhancing resilience and safety by improving forecasting and early warning systems, upgrading infrastructure, and conducting research and technology transfer for climate change adaptation.

2.5 In a Nutshell

This proposed strategy aims to provide a comprehensive framework for Vietnam's climate change adaptation efforts, demonstrating the country's commitment to addressing climate change. By prioritizing vulnerable people, regions, and sectors, mobilizing financial resources, and fostering whole-of-society engagement, Vietnam can safeguard its future against the unpredictable impacts of climate change. As one of the countries most affected by climate change, Vietnam must also seek international support to bolster its adaptive capacity. Stakeholders are encouraged to contribute feedback on this strategy to ensure it meets the needs of all affected parties.

References

Hallegatte S (2009) Strategies to adapt to an uncertain climate change. Glob Environ Change 19(2):240–247

Hallegatte S, Rentschler J, Rozenberg J (2020) The adaptation principles: a guide for designing strategies for climate change adaptation and resilience. World Bank, Washington, DC. https://openknowledge.worldbank.org/entities/publication/2dc19238-096a-5907-89c2-d4b99e6cb4d3

Ministry of Natural Resources and the Environment (2022) National adaptation plan (NAP) for the period 2021–2030, with a vision to 2050. https://www.undp.org/Vietnam/publications/national-adaptation-plan-period-2021-2030-vision-2050

Serrat O (2017a) From strategy to practice. In: Knowledge solutions: tools, methods, and approaches to drive organizational performance. Springer, pp 107–110

Serrat O (2017b) Creating and running partnerships. In: Knowledge solutions: tools, methods, and approaches to drive organizational performance. Springer, pp 7–10

Serrat O (2017c) Enriching policy with research. In: Knowledge solutions: tools, methods, and approaches to drive organizational performance. Springer, pp 937–944

Serrat O (2024) Redefining norms: a modern-day Marshall Plan for climate adaptation and climate justice. Unpublished manuscript, Georgetown University. https://www.researchgate.net/publication/380530326_Redefining_Norms_A_Modern-Day_Marshall_Plan_for_Climate_Adaptation_and_Climate_Justice

United Nations (2015) Transforming our world: the 2030 agenda for sustainable development. https://sdgs.un.org/goals

United Nations (2024) COP29. https://www.un.org/en/climatechange/cop29

World Bank (2022) Vietnam—Country climate and development report. Washington, DC.http://documents.worldbank.org/curated/en/099355507062229876/P17724103036ee07909701028a237b6d18b

World Resources Institute (2022) Locally led adaptation: from principles to practice. https://www.wri.org/research/locally-led-adaptation-principles-practice

Part II
Climate Change & International Security

Chapter 3
Beggar Thy Neighbor to Beggar Thyself: Why China and the United States Must Cooperate to Address Climate Change

Abstract Ideological differences, economic competition, security concerns, and disputes over human rights have sparked tense standoffs between China and the United States. Driven by geopolitical dynamics, the relationship between China and the United States will continue to evolve; even so, the two countries share outsized responsibility for climate change and their cooperation is crucial to achieving the 1.5°C temperature goal set out in the 2015 Paris Agreement. This précis suggests that by prioritizing bilateral cooperation, international forums, technological exchange, and climate change adaptation and resilience, senior officials in China and the United States can steer their nations to address pressing climate imperatives in a win–win partnership for greater common good.

3.1 Beggar Thy Neighbor to Beggar Thyself

In his inquiry into the nature and causes of the wealth of nations, Adam Smith regretted that mercantilism had motivated countries to beggar their neighbors (Smith 1776). Courtesy of Adam Smith, the term "beggar thy neighbor" refers to policies that a country enacts to adversely affect its trading partners, for example by means of protectionist barriers (e.g., quotas, sanctions, tariffs) or competitive devaluation; the basic idea is to reduce a nation's reliance on imports while increasing the demand for its exports. In contrast with the economic costs that "beggar-thy-neighbor" policies impose on others, "beggar-thyself" policies are those whose costs are borne primarily at home: they reflect domestic decisions to sacrifice economic efficiency to a competing social value (e.g., full employment, universal health care) or—in the worst case—miscalculations of domestic policy; the austerity that is eating away at Britain after Brexit comes to mind.

Never more than cordial, the relationship between China and the United States has deteriorated to the extent that it has become commonplace to characterize the intensifying competition between the two countries as "a new Cold War" (Marcus 2021). Climate change in the Anthropocene means that what goes around comes around: climate security rhymes with national security but also with international

O. Serrat, *Anthropogenic Solutions for Climate Change*,
https://doi.org/10.1007/978-981-96-8654-4_3

security because the two are now inseparably linked: to beggar others is to impoverish oneself (Serrat 2024). Because China and the United States are the world's largest carbon polluters, senior officials in the two countries can steer their nations to address climate imperatives in a universal win–win partnership. In 2022, China was the biggest emitter of carbon dioxide (CO_2) with about 29% of global emissions ahead of the United States (about 11%), but the United States is the biggest emitter in history (1750–2022), with more than twice China's emissions and the average American is responsible for nearly twice as much climate pollution as the average person in China (Statista 2024). This précis identifies cooperation opportunities for China and the United States to address climate change in four areas: (a) bilateral cooperation, (b) international forums, (c) technological exchange, and (d) climate change adaptation and resilience.

3.2 Back to the 1930s?

The 1930s are a byword for how bad things can get on the domestic front and in international relations: austerity, banking panics, economic nationalism, ideological clashes, isolationism, mass unemployment, protectionism, totalitarianism, war … Are we living through another 1930s? In the United States, the "Make America Great Again" (MAGA) slogan of Donald Trump's 2016 presidential campaign betrayed a genuine sense of disquiet about how the fruits of globalization's unsustainable boom had been divvied up to China's lop-sided benefit, notably after China became a member of the World Trade Organization in 2001 (Serrat 2000). In the wake of the 2007–2008 financial crisis, other shocks over the last 15 years (e.g., MAGA, Brexit, the COVID-19 pandemic, the Russian invasion of Ukraine) have shaken belief in market forces and endless progress. Born of geopolitical tensions, the markers of deglobalization are localization of global supply chains, reduced foreign direct investment, restrictive trade barriers, and more stringent regulation (Serrat 2023). The "beggar-thy neighbor" trade war that President Trump's administration instigated against China in 2018, and China responded in kind to, is reminiscent of the politics of anger of the 1930s. Around the world, governments are aware of the risks that climate change poses (Serrat 2012). Time is not on their side: to cap emissions of CO_2—the most dominant greenhouse gas produced by the burning of fossil fuels, industrial production, and land use change, governments know they must change the way their economies are run to find more socio-ecologically sustainable ways of life. (Other greenhouses gases include methane, nitrous oxide, ozone, and fluorinated gases, which have also contributed to global warming.) But change requires effort, resilience, planning, and support. For sure, change is not easy when the status quo is comfortable, the change is broad and vague, one is unclear about the benefits of change, many things must be changed at once, each change forces follow-on changes, results are slow to come, and we need others to change too (Tsaousides 2020). What is more, preoccupation with system maintenance will have a hegemon expect that others, not the hegemon itself, should change.

"Beggar thy neighbor" indeed. Extended in the arena of climate change, the listless and inappropriate dilettantism that "beggar-thy-neighbor" policies encourage at home will impose generalized "beggar-thyself" effects on the world at large if its largest polluters do not cooperate. China and the United States should not free-ride on the actions of other countries that fight climate change: paraphrasing Machiavelli (2003), they should make mistakes of ambition, not mistakes of sloth, by developing the strength to do bold things rather than the strength to suffer and make others grieve.

3.3 Tracking China–U.S. External Relations

The relationship between China and the United States is one of the most important and complex in the world (Hanhimäki 2017). The following lists key moments in the evolution of China–U.S. external relations, extracted from a timeline drawn by the Council on Foreign Relations (2024):

- **1949.** Mao Zedong establishes the People's Republic of China on October 1 after the Communists defeat the Nationalist government of Chiang Kai-shek. The United States, which had backed the Nationalists during the Second World War, supports Chiang Kai-shek's exiled government in Taiwan: this leads to limited relations with China.
- **1950.** The Korean War breaks out on June 25 when the Korean People's Army invades South Korea. The United Nations and the United States rush to defend South Korea. In October 25, China strikes when United States, United Nations, and South Korean troops approach the Chinese border. On October 6–7, China's People's Liberation Army crosses the Jinsha River to secure the "peaceful liberation of Tibet."
- **1954–1955.** The First Taiwan Strait Crisis breaks out in August when Chiang Kai-shek deploys thousands of troops to Kinmen and Matsu, prompting China's People's Liberation Army to shell the islands. In the spring of 1955, President Eisenhower considers using nuclear weapons against China's People's Liberation Army in Fujian Province. China and the United States hold talks in Geneva in September 1955 to prevent the escalation of future conflicts. China resumes its bombardment of Kinmen and Matsu in 1958.
- **1959.** A widespread uprising occurs in Lhasa on March 10, nine years after China asserted control over Tibet. The uprising leads to the expulsion of the Dalai Lama. China accuses the West (including the United States) of trying to encourage independence in Tibet. China's control over Tibet continues to reverberate throughout its efforts to become a world power.
- **1964.** China conducts its first test of an atomic bomb on October 16, joining the nuclear club amid China–U.S. friction over the escalating conflict in Vietnam.
- **1972.** In February, President Nixon meets Chinese leader Mao Zedong in Beijing and signs the Shanghai Communiqué. The Shanghai Communiqué sets the stage

for improved China–U.S. relations by allowing the two countries to discuss difficult issues, particularly Taiwan.
- **1989.** On June 3–4, China's People's Liberation Army clears Beijing's Tiananmen Square of protestors demanding democratic reforms and an end to corruption. The number of fatalities is estimated to range from several hundred to several thousand. The United States suspends military sales to China and freezes relations.
- **1972–2016.** From President Nixon's visit to China in February 1972 to Barack Obama's presidency in 2008–2016, the relations between China and the United States remain cordial. In 2001, China joins the World Trade Organization to signal its integration into the global economy; hoping for economic reforms, the United States supports China's accession but friction over its trade practices continues to escalate.

From the 2010s, the relationship between China and the United States has become strained on a par with China's rise to second-largest economy status: notwithstanding heightened trade disputes—with the United States experiencing an all-time high trade deficit with China of $295.5 billion in 2011, China has boosted military spending and engaged in territorial disputes and controversial land reclamation efforts in the South China Sea (Council on Foreign Relations 2024). In 2017, President Trump's first national security strategy charged China with pursuing economic aggression designed to weaken America (Sevastopulo 2017). In 2018, the Trump administration imposed sweeping tariffs on Chinese imports and China retaliated. On July 23, 2020, U.S. Secretary of State Michael Pompeo delivered at The Richard Nixon Presidential Library and Museum in Yorba Linda, California a speech titled "Communist China and the Free World's Future." Referring to President Nixon's overture to China in 1972, Pompeo declared that the era of engagement with the Chinese Communist Party was over, condemning China's aggressive moves in the South China Sea as well as the East China Sea, human rights abuses in Hong Kong and Xinjiang, intellectual property theft, and unfair trade practices (U.S. Department of State 2020).

Increasing distrust in the aftermath of the COVID-19 pandemic seems to have exacerbated the divide between China and the United States in a relationship that has never been closer or more complex since 1949 (Zhang 2020). On November 15, 2023, President Biden and Chinese leader Xi Jinping met in Woodside, California to produce deliverables on countering narcotics, military dialogue, renewable energy, and more; China and the United States are still seen to engage in a contest for power, aiming also to set the global rules of the game, but welcome guardrails have nevertheless been created (The White House 2023).

Wither are China–U.S. external relations bound? The United States wastes no opportunity to make its views known but it goes without saying that understanding China's worldviews matters too. Ten years ago, Henry Kissinger, the doyen of realist foreign policy practitioners, compared these protagonists:

> The American approach to policy is pragmatic; China's is conceptual. America has never had a powerful threatening neighbor; China has never been without a powerful adversary on

> its borders. Americans hold that every problem has a solution; Chinese think that each solution is an admission ticket to a new set of problems. Americans seek an outcome responding to immediate circumstances; Chinese concentrate on evolutionary change. Americans outline an agenda of practical "deliverable" items; Chinese set out general principles and analyze where they will lead. Chinese thinking is shaped in part by Communism but embraces a traditionally Chinese way of thought to an increasing extent; neither is intuitively familiar to Americans. (Kissinger 2014, p. 226)

3.4 China–U.S. Cooperation in the Context of Climate Policy

Notwithstanding their disputes, it is possible for China and the United States to avoid a zero-sum game and reshape their relations for the greater good with what Zhang (2020) termed "rationality, patience, wisdom, and creativity" (p. 1). With the dissolution of the Soviet Union in 1991, the United States became the only superpower; however, China's reach is now global and several middle powers (e.g., Australia, Brazil, the European Union, India, Japan, South Korea, Turkey) are becoming more influential (Ashford 2023). In a multipolar and more chaotic world that equates with diminishing relative power, China and the United States must avoid tropes of the "with us or against us" variety: with flexibility and openness, they should cooperate and leverage the power of allies for common benefit (Ashford 2023). Auspiciously, against the background of the United Nations Framework Convention on Climate Change (UNFCCC 1992) but also that of the 2030 Agenda for Sustainable Development (UN General Assembly 2015), climate policy offers opportunities for a reset of China–U.S. relations that meets the needs of the times. Key moments in recent China–U.S. cooperation in the context of climate policy include:

- **2014: Joint Climate Announcement.** On November 11, 2014, President Obama and Chinese leader Xi Jinping issue a Joint Announcement on Climate Change on the sidelines of the 2014 Asia-Pacific Economic Cooperation summit in Yanqi Lake, near Beijing, pledging to reduce CO_2 emissions (American Presidency Project 2014). Chinese leader Xi Jinping makes China's first promise to curb the growth of CO_2 emissions by 2030. President Obama sets a more ambitious target for U.S. emissions cutbacks.
- **2015: COP21.** At the COP21 climate summit in Paris, China and the United States play pivotal roles in shaping the Paris Agreement to reduce CO_2 emissions and limit global warming, anchored in their 2013 Joint Announcement on Climate Change.
- **2021: COP26.** On April 17, 2021, China and the United States release the U.S.–China Joint Statement Addressing the Climate Crisis (U.S. Department of State 2021a) to signify they share the goal of raising global climate ambition on mitigation, adaptation, and support on the road to COP 26 in Glasgow. On November 10, 2021, China and the United States issue the U.S.–China Joint Glasgow Declaration on Enhancing Climate Action in the 2020s (U.S. Department of

State 2021b). In the latter declaration, they recognize the urgency of the climate crisis, reaffirm their commitment to strengthening the Paris Agreement, agree to cooperate on reducing methane and CO_2 emissions, commit to cooperate on developing regulatory frameworks for decarbonization and electrification, and plan to convene a meeting in 2022 to focus on enhancing measurement and mitigation of methane (McGrath 2021; U.S. Department of State 2021b).

- **2023: Sunnylands Statement on Enhancing Cooperation to Address the Climate Crisis.** On November 14, 2023, U.S. Special Presidential Envoy for Climate John Kerry and Xie Zhenhua, his Chinese counterpart, emphasize in Sunnylands, California their nations' commitment to confront the climate crisis. Their statement reaffirms the pledges made in the U.S.–China Joint Glasgow Declaration on Enhancing Climate Action in the 2020s; commits to the effective implementation of the United Nations Framework Convention on Climate Change; commits likewise to the goals of the Paris Agreement; announces the operationalization of the Working Group on Enhancing Climate Action in the 2020s, focusing on energy transition, methane, circular economy, and deforestation; agrees to specific actions to reduce methane and other non-CO_2 greenhouse gas emissions, including implementing national methane action plans and cooperating on policy dialogue and technical solutions; and highlights the importance of international cooperation and the role of COP28 in addressing the climate crisis (U.S. Department of State 2023).
- **2024: The Green Energy Landscape.** Concerns have emerged over China's aggressive expansion in the green energy sector. On March 27, 2024, U.S. Treasury Secretary Janet Yellen signifies that a glut of heavily subsidized green technology exports from China is warping supply chains (Rappeport 2024a). Chinese President Xi Jinping responds that "China and the United States should become boosters for each other's development, not obstructions on each other" (Rappeport 2024a, para. 13). On April 5, 2024, U.S. Treasury Secretary Janet Yellen and Chinese Vice Premier He Lifeng met in Guangzhou, China at the start of a four-day state visit to address what the United States sees as China's industrial overcapacity, notably regarding exports of green energy technology that from state-driven investment threaten the electric vehicle and solar sectors that the United States has been trying to develop and the Biden administration is prepared to protect (Rappeport 2024b). U.S. Treasury Secretary Janet Yellen appealed to China for more balanced economic growth and trade practices as well as cooperation on climate change and other global challenges (Rappeport 2024b).

These recent developments confirm broader recognition of the importance of cooperation between China and the United States in addressing the global challenge of climate change. Even so, the rapport between the two countries is complex and evolving: persistent political tensions over trade practices, human rights, Taiwan, security in the Asian and Pacific region, and technology and cybersecurity overshadow dialogue; China and the United States are at different stages of development; China's economy is heavily dependent on coal, which provides 70% of its energy; and China is loath to be seen to bow to pressure from the United States (de la Garza 2023). At the 2024 Boao Forum for Asia that took place in Hainan Province on March 26–29, 2024 on the theme of "Asia and the World: Common Challenges, Shared Responsibilities," China's new climate envoy Liu Zhenmin summoned greater international cooperation for expansion of renewable energy globally.

3.5 Policy Recommendations

Problem statements help concerned parties focus on issues and solutions. The U.S.–China Joint Glasgow Declaration on Enhancing Climate Action in the 2020s can be interpreted as an acknowledgment that climate change is a significant global issue and that China and the United States have agreed to cooperate on actions to combat it. In the United States, several departments and agencies might be involved in international cooperation with China on climate policy. It is expected that the Department of State and the U.S. Agency for International Development would be key players in the coordination of U.S. government efforts to align international climate policy with national development priorities and engage in diplomacy to foster cooperation on climate change initiatives. The Senior Advisor to the President for Clean Energy Innovation and Implementation, in which capacity John Podesta currently serves, would also play a significant role in climate diplomacy and policy implementation, as seen with the Inflation Reduction Act of 2022. In China, it is expected that the Ministry of Foreign Affairs and the Ministry of Ecology and Environment would likewise be key players in the coordination of Chinese government efforts to align international climate policy with national development priorities and engage in diplomacy to foster cooperation on climate change initiatives. The Ministry of Science and Technology would also be involved, especially in the areas of scientific research and development related to climate change and clean energy technologies. Suggestions for measures that senior officials in China and the United States might take to achieve positive outcomes are in the box below. Toward each desired end, senior officials in China and the United States would need to recognize their respective domestic and international contexts; identify interests, threats, and opportunities; surface assumptions and biases; articulate specific, measurable, achievable, and timebound (SMART) objectives; identify (or develop) means; frame ways; and assess costs, risks, and results. Capacity for new behavior will be built if senior officials in China and the United States listen, respect, and suspend judgment before they voice (Isaacs 1999) Box 3.1 highlights key areas where China and the United States can collaborate to address climate change.

Box 3.1 Cooperation opportunities for China and the United States to address climate change

- Bilateral Cooperation
  - **Rationale:** Climate change is a universal leveller and cooperation to address it takes precedence over political rivalry.
  - **Recommendation:** Build trust for bilateral cooperation by forging, enhancing, and renewing shared commitments, such as the China–U.S. Joint Statement Addressing the Climate Crisis and the China–U.S. Joint Glasgow Declaration on Enhancing Climate Action in the 2020s.
  - **Recommendation:** Encourage and support subnational entities (e.g., cities, corporations, states) to share best practices, develop joint projects, and contribute to achieving climate targets including climate-friendly trade.
- International Forums
  - **Rationale:** With enhanced negotiation power, China and the United States can exercise global leadership to set a powerful example for other nations and encourage worldwide commitment to climate goals.
  - **Recommentation:** Strengthen multilateral cooperation: engage in climate dialogues, enhance transparency, and align policies.
  - **Recommendation:** Jointly advocate for ambitious climate targets at international fora.
- Technological Exchange
  - **Rationale:** China and the United States are world leaders in green technology: international cooperation can synergize innovation to accelerate breakthroughs in renewable energy and carbon capture and storage.
  - **Recommendation:** Create bilateral innovation hubs and encourage joint research centers to develop and commercialize green technology.
  - **Recommendation:** Facilitate technology transfer while safeguarding intellectual property rights.
- Climate Change Adaptation & Resilience
  - **Rationale:** Slow-onset events (e.g., desertification, drought, erosion, sea-level rise) and rapid-onset events (e.g., cyclones, flash floods, storms) affect both China and the United States.
  - **Recommendation:** Establish and advance a Bilateral Climate Adaptation and Resilience Dialogue.
  - **Recommendation:** Share best practices in climate change adaptation and resilience and collaborate on climate resilience strategies, planning, disaster response, and infrastructure upgrades.

3.6 Facing the Future

As the 2024 United States presidential election draws near, the China–U.S. relationship will be scrutinized.[1] Irrespective of the outcome of the 2024 United States presidential election, tensions over trade practices, human rights, Taiwan, security in the Asia-Pacific region, technology and cybersecurity, and China's recent alignment with Russia could upset the delicate balance of bilateral relations and limit opportunities for climate negotiation and cooperation. Nonetheless, the interplay between geopolitics and climate change is more pronounced than ever. Can the world's two largest emitters of CO_2 put aside their political differences and start working together to champion the radical lowering of global CO_2 emissions and address a common existential threat? Surely, everyone hopes so. In the immediate, the cooperation opportunities tabled in this précis could be considered during eventual consultations to renew the U.S.–China Science and Technology Cooperation Agreement that was instituted on January 31, 1979 under President Carter and Chinese leader Deng Xiaoping and expired on February 27, 2024 (U.S. Congressional Research Service 2023). The stated objective of the U.S.–China Science and Technology Cooperation Agreement is to provide opportunities for cooperation in science and technology fields of mutual interest, thereby promoting progress for the benefit of both countries, and consequently for the world at large, with cooperation based on equality, reciprocity, and mutual benefit.

3.7 The Time for Earnest China–U.S. Cooperation to Address Climate Change is Now

As the two largest emitters of CO_2 globally, China and the United States have a moral obligation to demonstrate climate leadership (Serrat 2017b); the clamor to "redefine climate change from a human rights and environmental justice perspective" is growing but mere self-interest surely justifies it (Corpwatch 2002, para. 1). Despite this, less than six months after his inauguration on January 20, 2017, President Trump signified in unapologetic "beggar-thyself" mode his intention to withdraw from the Paris Agreement, claiming that the economic burden it imposed on American workers, businesses, and taxpayers was unfair (U.S. Department of State 2019). The blowback to President Trump's announcement to exit the Paris climate agreement was thunderous, both in the United States and abroad; the *Berliner Kurier*, a tabloid, did not mince rude words (Guy 2017; Watts and Connolly 2017). On January 20, 2021, shortly after his inauguration, President Biden signed an executive order to rejoin the Paris Agreement (U.S. Department of State 2021c). Vitally, on August 16, 2022, President Biden signed the Inflation Reduction Act in

[1] *Beggar Thy Neighbor to Beggar Thyself: Why China and the United States Must Cooperate to Address Climate Change* was completed on April 30, 2024 ahead of the 2024 United States presidential election.

a comprehensive effort to drive climate action with the largest investment in clean energy in the history of the United States and to create jobs, reduce costs, and promote fairness in the economy (U.S. Department of the Treasury 2024). Even so, President Trump's decision blighted the reputation of the United States, with repercussions to this day.

Evading your tasks today will not free you from tomorrow's obligations: the United States must face the future or shortly face the music. At the press conference announcing the U.S.-China Joint Glasgow Declaration on Enhancing Climate Action in the 2020s, according to Shapiro et al. (2021), U.S. Special Presidential Envoy for Climate John Kerry—for one—acknowledged the moral responsibility of the United States to provide solutions to climate change given its history of contributing to the problem:

> And, yes, we have a fundamental moral obligation to do this [U.S. Special Presidential Envoy for Climate John Kerry said] because we are the richest country on the planet. We're the second-largest emitter, and we've been doing this for a long time. And the accumulated results of what we've been doing are up in the atmosphere causing damage, and we need to pay attention to that. (As cited in Shapiro et al. 2021, para. 24)

China and the United States are in a unique and critical position in the fight against climate change because their actions—or underperformances—have a disproportionate impact on the Earth's climate system. Climate change is a collective challenge that knows no borders and the logarithmic consequences of inaction have impact globally. China and the United States stand to gain economic, social, security, and reputational benefits from cooperative climate action that avoids "the tragedy of the commons" through enhanced efficiency and scientific advancement. Now is the time for earnest China–U.S. cooperation to address climate change.

References

American Presidency Project (2014) Joint statement—U.S.–China joint announcement on climate change, November 12. https://www.presidency.ucsb.edu/documents/joint-statement-us-china-joint-announcement-climate-change

Ashford E (2023) Yes, the world is multipolar. Foreign Policy, October 5. https://foreignpolicy.com/2023/10/05/usa-china-multipolar-bipolar-unipolar/

Corpwatch (2002) Bali principles of climate justice, August 28. https://www.corpwatch.org/article/bali-principles-climate-justice

Council on Foreign Relations (2024) 1943–2023: U.S.–China relations. https://www.cfr.org/timeline/us-china-relations

de la Garza A (2023) These two charts show why U.S.–China cooperation on climate is so challenging. Time, July 20. https://time.com/6295941/us-china-climate-cooperation-challenge/

Guy B (2017) Universal backlash from leaders on Trump's Paris blunder. Natural Resources Defense Council, June 9. https://www.nrdc.org/bio/brendan-guy/universal-backlash-leaders-trumps-paris-blunder

Hanhimäki J (2017) China and the United States: the evolution of a relationship. Global Challenges, 1. Geneva Graduate Institute. https://globalchallenges.ch/issue/1/china-and-the-united-states-the-evolution-of-a-relationship/

Isaacs W (1999) Dialogue: the art of thinking together. Doubleday Currency

Kissinger H (2014) World order. Penguin Press

Machiavelli N (2003) The prince (G. Bull, Trans.). Penguin Classics (Original work published 1532)

Marcus J (2021) U.S.–China relations: beyond the "cold war" cliche. British Broadcasting Corporation, March 16. https://www.bbc.com/news/world-asia-56382793

McGrath M (2021) COP26: China and U.S. agree to boost climate cooperation. British Broadcasting Corporation, November 10. https://www.bbc.com/news/science-environment-59238869

Rappeport A (2024a) Yellen warns China against flood of cheap green energy exports. New York Times, March 27. https://www.nytimes.com/2024/03/27/business/yellen-china-green-technology.html

Rappeport A (2024b) Yellen faces diplomatic test in urging China to curb green energy exports. New York Times, April 5. https://www.nytimes.com/2024/04/05/business/yellen-china-green-energy-exports.html

Serrat O (2000) China's economic prospects: a projection [PowerPoint slides]. https://www.researchgate.net/publication/360824361_China%27s_Economic_Prospects_A_Projection

Serrat O (2012) World sustainable development timeline. Asian Development Bank. https://www.adb.org/publications/world-sustainable-development-timeline

Serrat O (2017b) Moral courage in organizations. In: Knowledge solutions: tools, methods, and approaches to drive organizational performance. Springer, pp 489–497

Serrat O (2023) Globalization and deglobalization: must it be either/or? Unpublished manuscript, Georgetown University. https://www.researchgate.net/publication/374583890_Globalization_and_Deglobalization_Must_It_Be_EitherOr

Serrat O (2024) ICYMI: climate security rhymes with national security. Unpublished manuscript, Georgetown University. https://www.researchgate.net/publication/378545941_ICYMI_Climate_Security_Rhymes_with_National_Security

Sevastopulo D (2017) Trump labels China a strategic "competitor". Financial Times, December 18. https://www.ft.com/content/215cf8fa-e3cb-11e7-8b99-0191e45377ec

Shapiro A, Brown A, Caldwell N, Venkat M, Jarenwattananon P (2021) U.S. and China announce surprise climate agreement at COP26 summit. National Public Radio, November 11. https://www.npr.org/2021/11/11/1054648598/u-s-and-china-announce-surprise-climate-agreement-at-cop26-summit

Smith A (1776) An inquiry into the nature and causes of the wealth of nations. A. Strahan & T. Cadell

Statista (2024) Emissions. https://www.statista.com/markets/408/topic/949/emissions/#overview

The White House (2023) Readout of President Joe Biden's meeting with President Xi Jinping of the People's Republic of China, November 15. https://www.whitehouse.gov/briefing-room/statements-releases/2023/11/15/readout-of-president-joe-bidens-meeting-with-president-xi-jinping-of-the-peoples-republic-of-china-2/

Tsaousides T (2020) What makes change difficult? Approaches that make change hard and even impossible. Psychology Today, October 16. https://www.psychologytoday.com/us/blog/smashing-the-brainblocks/202010/what-makes-change-difficult

U.S. Congressional Research Service (2023) U.S.–China Science and Technology Cooperation Agreement, November 17. https://crsreports.congress.gov/product/pdf/IF/IF12510

U.S. Department of State (2019) On the U.S. withdrawal from the Paris Agreement, November 4. https://2017-2021.state.gov/on-the-u-s-withdrawal-from-the-paris-agreement/

U.S. Department of State (2020) Communist China and the free world's future, July 23. https://2017-2021.state.gov/communist-china-and-the-free-worlds-future/

U.S. Department of State. (2021a) U.S.-China joint statement addressing the climate crisis, April 17. https://www.state.gov/u-s-china-joint-statement-addressing-the-climate-crisis/

U.S. Department of State (2021b) U.S.-China joint Glasgow declaration on enhancing climate action in the 2020s, November 10. https://www.state.gov/u-s-china-joint-glasgow-declaration-on-enhancing-climate-action-in-the-2020s/

U.S. Department of State (2021c) The United States officially rejoins the Paris Agreement, February 19. https://www.state.gov/the-united-states-officially-rejoins-the-paris-agreement/

U.S. Department of State (2023) Sunnylands statement on enhancing cooperation to address the climate crisis, November 14. https://www.state.gov/sunnylands-statement-on-enhancing-cooperation-to-address-the-climate-crisis/

U.S. Department of the Treasury (2024) Inflation reduction act. https://home.treasury.gov/policy-issues/inflation-reduction-act

UN General Assembly (2015) Transforming our world: the 2030 agenda for sustainable development. A/RES/70/1. https://sdgs.un.org/

UNFCCC (1992) United Nations framework convention on climate change. https://unfccc.int/resource/docs/convkp/conveng.pdf

Watts J, Connolly K (2017). World leaders react after Trump rejects Paris climate deal. The Guardian, June 1. https://www.theguardian.com/environment/2017/jun/01/trump-withdraw-paris-climate-deal-world-leaders-react

Zhang B (2020) Understanding changes in Sino–U.S. relations from a historical perspective. China Int Strat Rev 2:1–13

Chapter 4
ICYMI: Climate Security Rhymes with National Security

Abstract Climate security rhymes with national security. However, the definition that the United States Congress has given to climate security means that—foremost among governments—the United States has paid significant attention to its military bases and overseas interests. This précis contends that true climate security necessitates a more expansive outlook that does not just serve blinkered national agendas. Climate security rhymes with national security but also with international security since the two are now inseparably linked.

The Last Judgment, an oil-on-wood triptych that Hieronymus Bosch (ca. 1450–1516) completed ca. 1482, depicts some of the most famous scenes of the underworld. Intended as a warning of the suffering in store for sinners, the right panel of *The Last Judgment* is a landscape of torture by fantastical demons. Bosch believed that hell was the upshot of humankind's ultimate sin: its attachment to pleasure. For "pleasure," 500-odd years later, read "arrogance," "ignorance," and "greed" in the wealthier countries, or about 50 years of consumer capitalism. Bosch's hell is around the corner.

By 2100, if we continue to do so little about climate change in the Anthropocene, global temperatures will have risen by more than 4°C above preindustrial levels, 1850–1900 being the designated reference period: heat waves will be triggered in many parts of the world, especially in the tropics; vast areas of the Earth will be experiencing droughts; the Arctic will be free of ice every summer; thermal expansion and melting ice sheets will have caused sea levels to rise by over 1 meter, inundating coastal cities; high-intensity storms and cyclones will have become commonplace; rainfall will be more seasonal and concentrated, with repeated flash flooding; the marine food chain will have collapsed in regions because of acidification and overfishing; food and water insecurity will plague the lives of billions of people; new pandemics will have started to occur; populations will have migrated to temperate zones; and, biodiversity including coral reef systems will have been lost (Maslin 2021). If you think that 80 years is plenty of time to buy a penthouse in Greenland, think again: bringing together the latest advances in climate science, the Sixth Assessment Report of the United Nations Framework Convention on Climate Change announced in 2021 that global warming will hit the 1.5°C threshold of the 2015 Paris Agreement as early as 2040; issued "a code red for humanity"; and urged

O. Serrat, *Anthropogenic Solutions for Climate Change*,
https://doi.org/10.1007/978-981-96-8654-4_4

immediate, rapid, and large-scale reductions in greenhouse gas emissions (UN News 2021). The United Nations Intergovernmental Panel on Climate Change's forecast may transpire sooner: in 2023, for the first time in recorded history, the global temperature pushed past the warming threshold for an entire year from February 2023 to January 2024, running 1.52°C (Copernicus 2024). There has been progress toward sustainable development but it is not nearly enough (Serrat 2012, 2023).

The word "religion" derives from the Latin verb "*religare*," meaning, "to fasten" or "to bind stronger"; accordingly, religion is the connection that links to divinity. Farsightedly, Dubos (1972) held that "[o]ur salvation depends upon our ability to create a religion of nature" (p. 41). This would seem intuitive: humankind is part of nature, not above it, and we have a human need to belong so we might be greater. In our climate predicament, we must be, think, and act bigger than our discrete selves. With decisive action, a 4°C world can be avoided and global warming can probably be held below 2°C: this would mitigate the bloodcurdling risks that will be induced by changes in climate patterns. Taking a leaf from Dubos (1972), our pragmatic religion must be that of climate security.

The connection between the effects of a changing climate and security, be it—broadly—the safety and well-being of human societies or—narrowly—menace to civil order, is clear. In view of that, why are we not addressing climate change forcefully? Markman (2018) had answers: first, acting on climate change represents a trade-off between short-term and long-term benefits, something we find hard to balance; second, climate change is a nonlinear, multidimensional problem; third, many effects of climate change seem far-off in time; fourth, the future is always less certain than the present. Still, Markman (2018) might have acknowledged that poor populations do not have much say anyway. The poor, not the better-off, are already staring the crisis in the face: they are most vulnerable to climate change because of high dependence on natural resources and limited capacity to cope with variability and extremes. Presumably addressing the more comfortable members of world society, Markman (2018) suggested things individuals might do, namely, bringing the future mentally close, confronting the uncertainty of the future head-on, and conducting serious discussions about values with one's peers and within one's organization. Ultimately, do people in wealthier countries want to live their lives without regard to the welfare of their children, grandchildren, and worthier others they will never know? "Are we being good ancestors?", Salk (1977) asked. If we cannot countenance selfish values, we must change our behavior.

We should not lowball the power of individuals to exercise values through choices. Besides, environmental ethics are indispensable part of effective solutions to climate change. All the same, we need traction now. Some charge that the nation-state fired up climate change by its efforts to boost the living standards of citizens; toward this, Friedman (1970) blessed corporate greed. Others think that a planetary issue such as climate change is too challenging to tackle in a world of nation-states. Then again, it may be the overarching concern that the nation-state has for national interest and territorial sovereignty, as well as the resources at its disposal, that equips it for climate action. The nation-state can be charged with being part of the problem; even so, like it or not, we live in a world system that—at least for the time being—retains capitalism as its base and the nation-state as its superstructure.

Brown (1977) reasoned that the threats to national security might soon arise less from the relationship of nation to nation and more from the relationship of man to nature. In the mid-2000s, climate change was labeled an "accelerant of instability," or a "threat multiplier," and climate security emerged as a framework for representing the economic, physical, and societal impacts of climate change that may affect peace and stability. Definitions of climate security vary in keeping with the actor. In recent legislation authorizing the Climate Security Advisory Council in 2020, the United States Congress defined climate security as:

- The national security of the United States, including national security infrastructure
- Subnational, national, and regional political stability
- The security of allies and partners of the United States
- Ongoing or potential political violence, including unrest, rioting, guerrilla warfare, insurgency, terrorism, rebellion, revolution, civil war, and interstate war. (50 U.S. Code § 3060 2021, para. 6)

De facto, the definition that the United States Congress gave to climate security means that—foremost among governments—the United States has paid highest attention to military bases and overseas interests. Critics have argued that the said definition encourages a militarized response to the climate crisis; in addition, it ignores the maldistribution and inequity that in fragile states exacerbate vulnerability to climate change. Note also the absence of causal factors in the same definition, that Busby (2022) identified as weak state capacity, exclusive political institutions, and blocked or unevenly distributed foreign assistance. Interpreting our common quandary differently, the United Nations Development Programme (2020), the United Nations' lead agency on international development, recommended that three risk dimensions should be weighed up when analyzing climate security:

- **Climate Stressors or Shocks.** Includes erratic, extreme and/or changed rainfall patterns, temperature increase, storms, shifting seasonal patterns, and ecosystem degradation)
- **Exposure.** The presence of people, livelihoods, natural resources, infrastructure, or economic, social, or cultural assets in places that could be adversely affected; [and]
- **Vulnerability or Coping Capacity.** The propensity of exposed elements to be adversely affected and the ability of systems to manage and overcome adverse conditions. (United Nations Development Programme 2020, p. 3)

And so, yes: climate security rhymes with national security. However, if it is to mobilize responses to the climate challenge, true climate security necessitates a more expansive outlook in an international coalition of the willing that does not just serve blinkered national agendas: beyond militarized—and necessarily localized—climate responses, this summons climate mitigation and climate adaptation worldwide. Who should foot the bill? Lower-income, lower-emission countries face the highest risks. In addition to cutting their colossal emissions of carbon dioxide, countries such as the People's Republic of China, the United States, India, the European Union, and Russia must from a justice perspective organize vastly increased access to and deployment of climate finance and technical support for capacity-constrained and at-risk countries. Cumulative emissions from fossil fuels, land use, and forestry over the period 1850–2021 evidence that the United States, the People's Republic of China, Russia, Brazil, and Indonesia, in descending order, bear historic responsibility for the largest share of historical emissions, the latter two on account of emissions from deforestation (Carbon Brief 2021). Climate security rhymes with national security but also with international security since the two are now inseparably linked.

References

50 U.S. Code § 3060 (2021) Climate Security Advisory Council. https://www.law.cornell.edu/uscode/text/50/3060

Brown L (1977) Redefining national security. In: Worldwatch paper, vol 14. Worldwatch Institute, pp 1–46

Busby J (2022) States and nature. Cambridge University Press

Carbon Brief (2021) Analysis: which countries are historically responsible for climate change?, October 5 https://www.carbonbrief.org/analysis-which-countries-are-historically-responsible-for-climate-change/

Copernicus (2024) Warmest December concludes warmest year on record, January 11. https://climate.copernicus.eu/warmest-december-concludes-warmest-year-record

Dubos R (1972) A god within: a positive philosophy for a more complete fulfillment of human potentials. Charles Scribner's Sons

Friedman M (1970) A Friedman doctrine: the social duty of enterprise is to increase its earnings. The New York Times Magazine, September 13. https://www.nytimes.com/1970/09/13/archives/a-friedman-doctrine-the-social-responsibility-of-business-is-to.html

Markman A (2018) Why people aren't motivated to address climate change. Harvard Business Review, October 11. https://hbr.org/2018/10/why-people-arent-motivated-to-address-climate-change

Maslin M (2021) How to save our planet: the facts. Penguin Life

Salk J (1977) Are we being good ancestors? Acceptance Address for Jawaharlal Nehru Award for International Understanding, New Delhi, India, January 10.

Serrat O (2012) World sustainable development timeline. Asian Development Bank

Serrat O (2023) Localizing climate action in Latin American and Caribbean cities. Unpublished manuscript, Georgetown University. https://www.researchgate.net/publication/376032182_Localizing_Climate_Action_in_Latin_American_and_Caribbean_Cities

UN News (2021) IPCC report: "Code red" for human driven global heating, warns UN chief, August 9. https://news.un.org/en/story/2021/08/1097362

United Nations Development Programme (2020) Climate security mechanism: conceptual approach. New York.

Chapter 5
Redefining Norms: A Modern-Day Marshall Plan for Climate Adaptation and Climate Justice

Abstract With global impact, climate leadership has emerged as a new and critical area in external relations that reflects the interconnected nature of the climate crisis. Leading in climate action enhances a country's reputation on the world stage and helps it set international norms and expectations. This précis articulates an ambitious climate adaptation and climate justice strategy for the United States, to be presented as a modern-day Marshall Plan in Baku, Azerbaijan at the 2024 United Nations Climate Change Conference (COP29).

5.1 In a Word[1]

This précis articulates an ambitious climate adaptation and climate justice strategy for the United States, to be presented on November 11–22, 2024 at the United Nations Climate Change Conference (COP29) in Baku, Azerbaijan. With "whole-of-government" and "whole-of-society" approaches orchestrated by the Office of the Special Presidential Envoy for Climate, the strategy would galvanize action at home and abroad to deliver long-awaited climate adaptation and climate justice in the Anthropocene.

5.2 Introduction

Upon assuming office, President Biden issued Executive Order No. 13990, *Protecting Public Health and the Environment and Restoring Science to Tackle the Climate Crisis* (The White House 2021a) and Executive Order No. 14008, *Tackling*

[1] *Redefining Norms: A Modern-Day Marshall Plan for Climate Adaptation and Climate Justice* was completed on May 12, 2024 ahead of the 2024 United States presidential election. The fundamental guidance offered by this précis transcends individual candidates; it retains its relevance regardless of who eventually wins the election.

O. Serrat, *Anthropogenic Solutions for Climate Change*,
https://doi.org/10.1007/978-981-96-8654-4_5

the Climate Crisis at Home and Abroad (The White House 2021b). President Biden reiterated that climate change is a clear and present danger to the United States. Yet, President Biden also affirmed that the climate crisis is an opportunity for cost savings, energy efficiency, green economy expansion, innovation and research, health and productivity, investment attraction, job creation, market opportunities, resilience, and risk reduction. With climate action, economic reforms, and social benefits, the Inflation Reduction Act of 2022 marked one of the largest investments the Congress of the United States has made (U.S. Department of the Treasury 2024). Federal agencies have carried out climate-related executive orders and initiatives of President Biden's administration. Even so, the United States must intensify its efforts to tackle the climate crisis abroad so it might better protect and sustain Americans at home. Climate leadership is a new and critical area in external relations: by leveraging state-sponsored altruism, the United States can export its values and brand of organized capitalism, set norms and expectations, and enhance its reputation.

5.3 The Setting: Baku, Azerbaijan

Much as the earlier Conferences of the Parties, COP29 plans to advance the goals of the United Nations Framework Convention on Climate Change, especially those of the Paris Agreement of December 12, 2015, to "[hold] the increase in the global average temperature to well below 2°C above pre-industrial levels" and "limit the temperature increase to 1.5°C above pre-industrial levels" (UNFCCC 2024, para. 2). COP29 means to:

- **Limit Global Warming.** COP29 will endeavor to strengthen commitments to limit global warming by implementing ambitious climate action plans: this includes revisiting nationally determined contributions and enhancing emission reduction targets.
- **Adapt to Climate Change.** COP29 will address strategies for adapting to the impacts of climate change: this involves developing resilience measures, especially for vulnerable communities and ecosystems.
- **Mobilize Climate Finance.** COP29 will seek to mobilize financing for climate-related activities: this concerns funding goals, funding mechanisms, transparency, and accountability in climate finance.

To hasten the accomplishment of its objectives, COP29 proposes to enhance the transparency framework—so that countries might report on climate actions transparently and consistently—and emphasize equity and justice.

Emissions of carbon dioxide (CO_2) must be reduced. The United States is the largest emitter of CO_2 in history (1750–2022) with more than twice China's emissions (Statista 2024). In 2021, President Biden set a new target for the United States to achieve a 50–52% reduction from 2005 levels in economy-wide net greenhouse gas pollution in 2030 (The White House 2021c). COP29 will take place in the immediate aftermath of the presidential election in the United States: ahead of that, President Biden would earn plaudits at home and abroad with a decision to fast-track climate adaptation and climate justice. Paraphrasing Thunberg (2019), President Biden would have acted in time.

5.4 A Modern-Day Marshall Plan for Climate Adaptation and Climate Justice

The Marshall Plan, formally the European Recovery Program, was enacted in 1948 to restore the economies of 16 western and southern European countries after the devastation of the Second World War. Offering hope for those who needed it, the Marshall Plan challenged European leaders to cooperate and coordinate by drafting programs for reconstruction and development that the United States would then fund (Roll 2019). An exemplar of economic statecraft and international cooperation, the Marshall Plan increased agricultural and industrial production, liberalized trade, provided markets for American goods, strengthened transatlantic relations in an alliance that soon established the North Atlantic Treaty Organization, and curbed communist influence. Under the Marshall Plan, the United States transferred about \$13.3 billion (equivalent to \$172 billion in 2024) over the period 1948–1951. As it prepares for COP29, President Biden's administration can draw inspiration from the Marshall Plan to build a more resilient world that transcends our lifetimes.

5.4.1 The Strategic Situation

Climate change affects ecosystems, economies, and human lives and poses an unprecedented threat to our planet. International and domestic contexts are aligning: it is now understood that climate change also affects the local, threatening vital national interests including security and stability, infrastructure and resilience, public health and well-being, economic prosperity and trade, and environmental and cultural heritage (Serrat 2024a, b). Salient opportunities to engage in climate adaptation and climate justice have to do with evidence-based strategies, education and training, community participation, the building of resilient foundations with inclusive development, and investment in resilience measures. The United States is a major emitter of greenhouse gases and has a global responsibility to support climate adaptation and climate justice at home and abroad. The key question is:

> As the world grapples with extreme weather events, rising temperatures, and sea-level rise, is the United States willing to recognize its global responsibility and play a lead role in climate adaptation and climate justice?

In the affirmative, the climate adaptation efforts of the United States should prioritize justice, equity, diversity, and inclusion or economic and social disparities will widen. Two unknowns—but not unknowables—would relate to the ability to address cross-sectoral impacts from a system-wide perspective and to balance climate adaptation with climate justice (Serrat 2017a).

5.4.2 *The Desired End*

Most countries grapple with the escalating impacts of climate change. Therefore, enhancing the resilience and preparedness of communities, ecosystems, and infrastructure to withstand climate impacts is the desired end. Toward that, the specific objectives of a Modern-Day Marshall Plan for Climate Adaptation and Climate Justice should be to (a) embed climate resilience into planning and management, (b) invest in resilient infrastructure and frontline communities, and (c) extend financial and technical assistance toward the same to countries seeking help.

5.4.3 *The Means*

Ten components would be brought to bear:

- **Climate Adaptation.** Investments should be made in infrastructure and systems, e.g., sea walls, early warning systems, that increase resilience to climate impacts.
- **Climate Justice.** Policies, programs, and projects that protect vulnerable populations and promote social equity should be formulated.
- **Ecosystem Restoration.** Efforts should be made to restore and protect ecosystems that provide critical climate services, such as carbon sequestration and flood protection.
- **Financial Commitment.** Significant financial resources should be committed to support developing countries in their efforts to adapt to climate change and transition to green economies through such funding mechanisms as the Adaptation Fund, the Global Environment Facility, and the Green Climate Fund.
- **Global Cooperation.** A Modern-Day Marshall Plan for Climate Adaptation and Climate Justice would encourage international cooperation and coordination to tackle climate change.
- **Research and Development.** Increased funding for research and development can lead to new technologies and strategies that help adapt to climate change.
- **Technology Transfer.** Developed countries should share clean energy technologies, carbon capture and storage technologies, and climate-resilient agricultural practices with developing countries.
- **Policy and Regulatory Reform.** Support should be provided for policy and regulatory reform that promotes climate adaptation and mitigation, such as carbon pricing and renewable energy standards.

- **Capacity Building.** Assistance should be extended to developing countries so they might build the institutional and human capacity needed to effectively formulate and implement climate adaptation strategies.
- **Public Awareness and Education.** Resources should be dedicated to raising public awareness about climate change and promoting education in climate science and sustainable practices.

5.4.4 The Ways

Five cross-cutting priorities would underpin a commitment to climate adaptation:

- **Recognizing Climate-Related Economic and Financial Risks.** By recognizing, measuring, and disclosing economic and financial risks from climate change, a Modern-Day Marshall Plan for Climate Adaptation and Climate Justice would mitigate climate-related shocks at home and abroad.
- **Promoting Resilient Infrastructure.** A Modern-Day Marshall Plan for Climate Adaptation and Climate Justice would design and construct infrastructure that can withstand climate impacts, and extend assistance to countries seeking it.
- **Improving Community Resilience Planning.** At home and abroad, a Modern-Day Marshall Plan for Climate Adaptation and Climate Justice would promote collaboration with local communities to enhance their capacity to withstand climate-related shocks, for example by investing in early warning systems, community-based strategies, and infrastructure.
- **Conserving and Restoring Land and Water.** A Modern-Day Marshall Plan for Climate Adaptation and Climate Justice would prioritize conservation efforts, reforestation, wetland restoration, and sustainable land management, and extend assistance to countries seeking it.
- **Advancing Resilience Solutions.** At home and abroad, a Modern-Day Marshall Plan for Climate Adaptation and Climate Justice would support policies, research, and technology that enhance resilience across sectors.

Four cross-cutting priorities would underpin a commitment to climate justice:

- **Transitioning to Clean Energy.** A Modern-Day Marshall Plan for Climate Adaptation and Climate Justice would commit to a just transition away from fossil fuels: at home and abroad, this would include ensuring retraining and job opportunities for displaced workers when coal-fired power stations are retired, leveling the playing field for renewable energy technologies, and improving global access to components and materials.
- **Creating Climate-Friendly Jobs.** A Modern-Day Marshall Plan for Climate Adaptation and Climate Justice would create jobs in renewable energy, green infrastructure, and sustainable agriculture: these jobs should benefit all, especially those who have been left behind.

- **Investing in Frontline Communities.** Historically marginalized communities face disproportionate risks from climate change: a Modern-Day Marshall Plan for Climate Adaptation and Climate Justice would invest in their resilience, health, and well-being.
- **Displaying Good Governance.** Transparency, accountability, and inclusivity would guide all actions under a Modern-Day Marshall Plan for Climate Adaptation and Climate Justice. A Modern-Day Marshall Plan for Climate Adaptation and Climate Justice would champion collaboration with international partners, nongovernmental organizations, civil society, and researchers in academia.

The agencies tasked with linking means to ways through their elements of power, instruments, and interrelationships would be the Department of Agriculture, the Department of Energy, the Department of the Interior, the Department of Justice, the Department of State, the Environmental Protection Agency, the National Aeronautics and Space Administration, the National Oceanic and Atmospheric Administration, the United States Agency for International Development, and the United States International Development Finance Corporation.

5.4.5 The Costs, Risks, and Results

In today's money, the Marshall Plan cost about $158 billion over a four-year period. In April 2024, President Biden signed a $95 billion assistance package for Ukraine, Israel, and Taiwan, of which more than $60 billion for support to Ukraine. "It's going to make America safer. It's going to make the world safer. And it continues America's leadership in the world," President Biden explained (U.S. Department of Defense 2024, para. 1). The latest assistance to Ukraine added to $74.3 billion already sent by the United States between January 2022 and January 2024 (Masters and Merrow 2024). A Modern-Day Marshall Plan for Climate Adaptation and Climate Justice would make America and the world even safer and demonstrate yet more leadership. The annual cost of climate adaptation in developing countries is expected to range from $160 billion to $340 billion by 2030 and from $315 billion to $565 billion by 2050 (McKinsey and Company 2023). What the United States has expended for the defense of Ukraine is in the lower range of what will have to be spent on climate adaptation by developing countries in 2030. "The climate crisis is our third world war. It needs a bold response," Stiglitz (2019) headlined. The risks to a Modern-Day Marshall Plan for Climate Adaptation and Climate Justice are negligible: its results would be easily tracked and measured and would correlate closely with the intensity of efforts.

5.5 *Dum Spiro Spero*: Moving Forward

Climate adaptation is not an alternative for delayed action on climate mitigation: it is one element of a larger strategy that includes emissions abatement, CO_2 removal, and—conceivably—sunlight reflection. A Modern-Day Marshall Plan for Climate Adaptation and Climate Justice is acceptable, desirable, feasible, suitable, sustainable, and synergistic: its benefits in terms of avoided climate impacts, improved public health, and economic opportunities would be enormous. If a Modern-Day Marshall Plan for Climate Adaptation and Climate Justice is approved, China, the European Union, India, and other large emitters including Brazil, Indonesia, Japan, and Russia should be invited to join forces in a coalition of the willing that helps navigate a Modern-Day Marshall Plan for Climate Adaptation and Climate Justice with iterative course corrections (Serrat 2017b, c, d).

References

Masters J, Merrow W (2024) How much aid has the U.S. sent Ukraine? Here are six charts. Council on Foreign Relations, February 23. https://www.cfr.org/article/how-much-aid-has-us-sent-ukraine-here-are-six-charts

McKinsey & Company (2023) The cost of climate adaptation, November 27. https://www.mckinsey.com/featured-insights/sustainable-inclusive-growth/chart-of-the-day/the-cost-of-climate-adaptation

Roll D (2019) George Marshall: defender of the republic. Dutton Caliber

Serrat O (2017a) Understanding complexity. In: Knowledge solutions: tools, methods, and approaches to drive organizational performance. Springer, pp 345–353

Serrat O (2017b) Creating and running partnerships. In: Knowledge solutions: tools, methods, and approaches to drive organizational performance. Springer, pp 7–10

Serrat O (2017c) From strategy to practice. In: Knowledge solutions: tools, methods, and approaches to drive organizational performance. Springer, pp 107–110

Serrat O (2017d) Learning in strategic alliances. In: Knowledge solutions: tools, methods, and approaches to drive organizational performance. Springer, pp 639–647

Serrat O (2024a) ICYMI: climate security rhymes with national security. Unpublished manuscript, Georgetown University. https://www.researchgate.net/publication/378545941_ICYMI_Climate_Security_Rhymes_with_National_Security

Serrat O (2024b) Beggar thy neighbor to beggar thyself: why China and the United States must cooperate to address climate change. Unpublished manuscript, Georgetown University. https://www.researchgate.net/publication/380212272_Beggar_Thy_Neighbor_to_Beggar_Thyself_Why_China_and_the_United_States_Must_Cooperate_to_Address_Climate_Change

Statista (2024) Emissions. https://www.statista.com/markets/408/topic/949/emissions/#overview

Stiglitz J (2019) The climate crisis is our third world war. It needs a bold response. The Guardian, June 4. https://www.theguardian.com/commentisfree/2019/jun/04/climate-change-world-war-iii-green-new-deal

The White House (2021a) Executive order on protecting public health and the environment and restoring science to tackle the climate crisis, January 20. https://www.whitehouse.gov/briefing-room/presidential-actions/2021/01/20/executive-order-protecting-public-health-and-environment-and-restoring-science-to-tackle-climate-crisis/

The White House (2021b) Executive order on tackling the climate crisis at home and abroad, January 27. https://www.whitehouse.gov/briefing-room/presidential-actions/2021/01/27/executive-order-on-tackling-the-climate-crisis-at-home-and-abroad/

The White House (2021c) President Biden sets 2030 greenhouse gas pollution reduction target aimed at creating good-paying union jobs and securing U.S. leadership on clean energy technologies, April 22. https://www.whitehouse.gov/briefing-room/statements-releases/2021/04/22/fact-sheet-president-biden-sets-2030-greenhouse-gas-pollution-reduction-target-aimed-at-creating-good-paying-union-jobs-and-securing-u-s-leadership-on-clean-energy-technologies/

Thunberg G (2019) "You did not act in time": Greta Thunberg's full speech to MPs. The Guardian, April 23. https://www.theguardian.com/environment/2019/apr/23/greta-thunberg-full-speech-to-mps-you-did-not-act-in-time

U.S. Department of Defense (2024) Supplemental bill becomes law, provides billions in aid for Ukraine, Israel, Taiwan, April 24. https://www.defense.gov/News/News-Stories/Article/Article/3754718/supplemental-bill-becomes-law-provides-billions-in-aid-for-ukraine-israel-taiwan/

U.S. Department of the Treasury (2024) Inflation reduction act. https://home.treasury.gov/policy-issues/inflation-reduction-act

UNFCCC (2024) The Paris agreement. https://unfccc.int/process-and-meetings/the-paris-agreement

Part III
Deglobalization & Public Diplomacy

Chapter 6
Globalization and Deglobalization: Must It Be Either/Or?

Abstract This précis notes escalating interest in "deglobalization," a movement that owes to the backlash against globalization. All the same, the précis argues that globalization and deglobalization are work-in-progress and that consideration of the movements need not play one against the either. Considering the challenges the world faces, the précis suggests that deglobalization had best be treated as part of globalization's evolution, with implications for the solving of public problems, the restoration of trust, and the improved conduct of public diplomacy.

6.1 Terminology Exercises Hearts and Minds

In the volatile, uncertain, complex, and ambiguous (VUCA) world of the twenty-first century, we debate the meaning and significance of new words, disagree over definitions, sooner or later bump up newcomers in dictionaries, document their use in the real world, revise our descriptions, and sporadically remove them from our collections. Newfangled terms that capture the specificity of each happening are par for the course. In the field of politics and current events, an online dictionary recently registered "burnt pit," "cakeism," "ecofascism," "family office," "forever chemicals," "microtransaction," "self-coup," and "woke" (Dictionary.com 2023). In the realm of international relations, Haass (2017) proposed that "boots on the ground," "engagement," "fair trade," "freedom fighter," "global citizen," "international community," "liberal democratic order," "Middle East peace process," "right to self-determination," "smart power," "strategic dialogue," "superpower," and "surgical strike" should be taken to the woodshed. The generation of terminology has been stimulated by the online culture of our times, by those wishing to spotlight something, and by opponents who fight back.

O. Serrat, *Anthropogenic Solutions for Climate Change*,
https://doi.org/10.1007/978-981-96-8654-4_6

6.2 When Buzzwords Appeal

Sometimes, words and phrases are coined and gain currency because they do the job. To wit, in the sphere of international economics, this with cultural, economic, social, technological, trade, and consequently political dimensions, the term "deglobalization" has become prominent. Pryke (2022) traced the term to Bello (2002), which means that it is not new. Vitally, Pryke (2022) emphasized that deglobalization is a process, not an endpoint. Because the twenty-first century calls for interventions across a fast-moving economic, environmental, geopolitical, societal, and technological landscape, anything that aims to describe global processes makes for contested meanings hence terminological success. Google Trends (n.d.), a website by Google that since January 2004 analyzes the popularity of top search queries in Google Search, reveals growing worldwide interest in "deglobalization" as a topic beginning 2015–2016. Proof positive of the contemporary relevance of the topic of "deglobalization" is that, according to Google Trends (n.d.), it scored the maximum interest of 100 points in October 2011: that month was marked by the global protests of October 15, 2011, triggered by corporate influence over government, economic inequality, and political corruption, and conducted under the slogan "United for #GlobalChange."

6.3 Quid Globalization?

How one might define deglobalization depends on how one defines globalization. An indeterminate definition of globalization would connote the exchange of information, knowledge, goods, services, and ideas worldwide; that might serve to characterize the process of globalization up to the eighteenth century. Sachs (2020), however, referred to seven distinct ages: (i) the Paleolithic Age (70000–10000 BCE), (ii) the Neolithic Age (10000–3000 BCE), (iii) the Equestrian Age (3000–1000 BCE), (iv) the Classical Age (1000 BCE–1500 CE), (v) the Ocean Age (1500–1800 CE), (vi) the Industrial Age (1800–2000 CE), and (vii) the Digital Age (twenty-first century). Clearly, the seven stages in Sachs (2020) demonstrate intensification of process and extension of reach, that Merriam-Webster (2023) equated with "the development of an increasingly integrated global economy marked especially by free trade, free flow of capital, and the tapping of cheaper foreign labor markets." The annual meetings of celebrities, business leaders, economists, investors, journalists, policy makers, and political leaders that the World Economic Forum has held in Davos, Switzerland since 1971 were for long celebrations of free trade *joie de vivre* founded in utopian values of freedom, opportunity, and prosperity for every citizen. Note that Merriam-Webster's (2023) definition made no reference to impact. And yet, the 2007–2008 financial crisis, which had to do with excessive risk-taking by global financial institutions, led to the most severe worldwide economic crisis since the Great Depression (1929–ca. 1939); and so, beyond definitions of globalization as a process, discussions of the topic of globalization began to refer to its impact around the world. Brittanica (2023) alluded to both process and impact by characterizing globalization as "[A] long-term process of deterritorialization, that is, of social activities (economic, political, and cultural) occurring without regard for geographic location. Thus, globalization can be defined as the stretching of economic, political, and social relationships in space and time" (para. 2). Despite the foregoing, it is not that globalization is in decline: rather, as van der Marel (2020) pointed out:

> Although the COVID-19 crisis has seen a dramatic decline in goods trade, investments, and the movement of people, a new type of globalization is emerging. This "new globalization" is based on digital services, research and development, data, ideas, and other intangibles. (van der Marel 2020, p. 1)

If so, globalization may be becoming more intangible: "Globalization isn't dying: it's actually prospering albeit in different ways" (van der Marel 2020, p. 8). Van der Marel's (2020) assertion made intuitive sense. Contrary to Friedman's (2005) thesis, it is not that the world is flat. And, it is not that the world is spiky, per Florida's (2005) rejoinder. Wiggins, interviewed by Smith (2022) would agree with van der Marel (2020): "[Y]ou could argue [that] global trade will continue. Don't take this too far, right? Let's not call [this] the end of a world of global trade. It's just the end of the type of global trade that most people have got used to over three decades or more" (5:30 mins). I considered van der Marel (2020) an invitation to eschew binarism and consider globalization's subtle yin–yang relationship with deglobalization, its coevolutionary converse.

6.4 Has Deglobalization Come of Age?

The data from Google Trends (n.d.) cited earlier confirms that, on the heels of the 2007–2008 financial crisis, the debate over deglobalization has become more heated in consequence of economic, environmental, geopolitical, societal, and technological developments. The 2007–2008 financial crisis raised questions about the benefits of globalization and many governments in the Western world began to adopt protectionist policies. In early January 2017, Donald Trump was elected president of the United States on a platform to "Make America Great Again." The COVID-19 pandemic that broke out in late 2019 intensified the scrutiny of globalization. From that time, especially in Europe, right-wing populists began to score successes in elections. In late January 2020, the United Kingdom withdrew from the European Union, promising to "Take Back Control." Then, in early 2022, Russia invaded Ukraine in a sudden escalation of the Russo-Ukrainian War that had begun in 2014, trusting that Europe's reliance on Russian oil and natural gas would forestall opposition from the Western world. Put simply, the shocks of the last 15 years have shaken belief in market forces and endless progress. "In response to COVID-19, Russia's war with Ukraine, and climate change, governments and global companies are seeking security and resilience over the benefits of global value chains," Keller and Marold (2023) asserted.

Pryke (2022) advanced that, what with globalization having contested meanings, the term "deglobalization" cannot but hold limited appeal. Pryke's (2022) comment may have been premature: the term "globalization" has been a moving target, as we saw, but Britannica's (2023) definition made for clearer representation of, and consensus over, both process and impact; the same applies to the term "deglobalization." To Bello (2002), deglobalization was about reorienting economies from their emphasis on exports toward production for local markets. Then again, the term "deglobalization" can be expected to increasingly embody concern for national security and not merely reflect decreasing trends of import share in a country's gross domestic product. The prefix "de" smacks of generic opposition. More constructively, considering the challenges the world faces, deglobalization had best be treated as part of globalization's evolution. Global problems need global answers.

6.5 Leveraging Globalization and Deglobalization for Sustainability

We are hard-wired to worry—call it a survival technique—but we can surely learn to calm down. Helpfully, Harari (2018) suggested we should know which things to worry about and how much to worry about them. The first part is easy: right now, not in 2050, we face ecological, nuclear, and technological challenges; eight billion other persons will add to that list. Be that as it may, we can only tackle these pressing challenges as a global community: nationalist politics are irrelevant but fuel the fire. "We must, indeed, all hang together, or most assuredly we shall all hang separately," Benjamin Franklin is alleged to have said at the Continental Congress in 1776. Accepting that globalization and deglobalization are interconnected but not unavoidably opposite forces is a more pragmatic way to determine how much we should worry about them. Rephrasing Einstein's *bon mot*, Noveck (2021) stated the obvious: today's problems cannot be solved with yesterday's tools. In the Digital Age, Noveck (2021) recommended that the next generation of leaders and problem solvers should take advantage of digital technology, data, and the collective wisdom of communities to open governance and design and deliver powerful solutions. Still, how are we to emulate Benjamin Franklin's resolution? Trust is the foundation of strong relationships: it begets accomplishments. Deglobalization is an expression of mistrust in a fractious world. Accordingly, as stated by Manor and Surowiec (2021), "[T]he practice of public diplomacy must alter to contend with the emergence of uncertainty" (para. 1). To that end, away from propaganda as the simplest expression of so-called "external relations" or "foreign affairs," Cull (2019) identified discrete ways in which new-age public diplomacy can engage foreign publics: they are listening, advocacy, cultural diplomacy, exchange diplomacy, and news/international broadcasting. By any measure, integrated delivery of public diplomacy would be real progress.

References

Bello W (2002) Deglobalization: ideas for a new world economy. Zed Books

Brittanica (2023) Globalization. https://www.britannica.com/money/topic/globalization

Cull N (2019) Public diplomacy: foundations for global engagement in the digital age. Polity Press

Dictionary.com (2023) Winter 2023 new words: "Everything, everywhere, all at once", February 28. https://www.dictionary.com/e/new-dictionary-words-winter-2023/

Florida R (2005) The world is spiky: globalization has changed the economic playing field, but hasn't leveled it. Atl Mon 296(3):48–51

Friedman T (2005) The world is flat: a brief history of the twenty-first century. Farrar, Straus, & Giroux

Google Trends (n.d.) Google Trends. Retrieved September 22, 2023, from https://trends.google.com/trends/

Haas R (2017) 13 international relations buzzwords that need to get taken to the woodshed. Foreign Policy, February 3. https://foreignpolicy.com/2017/02/03/13-international-relations-buzzwords-that-need-to-get-taken-to-the-woodshed/#cookie_message_anchor

Harari, Y. (2018). 21 lessons for the 21st century. Vintage Books.

Keller C, Marold R (2023) Deglobalization: what you need to know. World Economic Forum, January 17. https://www.weforum.org/agenda/2023/01/deglobalisation-what-you-need-to-know-wef23/

Manor I, Surowiec P (2021) Public diplomacy and the politics of uncertainty, February 25. https://uscpublicdiplomacy.org/blog/public-diplomacy-and-politics-uncertainty

Merriam-Webster (2023) Globalization. https://www.merriam-webster.com/dictionary/globalization

Noveck B (2021) Solving public problems: a practical guide to fix our government and change our world. Yale University Press

Pryke S (2022) Deglobalization—What's in a world word? Global Policy, July 27. https://www.globalpolicyjournal.com/blog/27/07/2022/deglobalisation-whats-world-word

Sachs J (2020) The age of globalization: geography, technology, and institutions. Columbia University Press

Smith J [Host] (2022) The end of globalization at Davos. FT News Briefing [Audio podcast episode], May 24. https://www.ft.com/content/69057969-a0e3-4af0-9f34-89e6ff15476b

Van der Marel E (2020) Globalization isn't in decline: it's changing. European Centre for International Political Economy. ECIPE Policy Brief No. 6., 1–10. https://ecipe.org/publications/globalization-is-changing/

Chapter 7
Localizing Climate Action in Latin American and Caribbean Cities

Abstract Focusing on local climate action, this précis underscores the benefits of a systems approach to the delivery and maintenance of sustainable urban infrastructure. Specifically, the précis identifies and submits to subnational actors (viz., mayors and city managers) five recommendations to promote an energy—housing—mobility framework through local and practical pathways: (a) make more widespread use of clean energy; (b) raise energy efficiency; (c) increase and decentralize the number of clean producers; (d) improve demand management; and (e) hone strategies for planning, designing, and building. The précis also showcases multilevel governance arrangements and pilot programs and projects that Latin American and Caribbean cities have evolved, developed, and rolled out.

7.1 Policy Summary

Recommendations to localize climate action in Latin American and Caribbean cities

- Make more widespread use of clean energy.
- Raise energy efficiency.
- Increase and decentralize the number of clean producers.
- Improve demand management.
- Hone strategies for planning, designing, and building.

Focusing on local climate action, this précis underscores the benefits of a systems approach to the delivery and maintenance of sustainable urban infrastructure. The précis identifies and submits five recommendations to promote an energy—housing—mobility framework through local and practical pathways. The recommendations are addressed to subnational actors (viz., mayors and city managers), who would have to engage non-state stakeholders (e.g., local communities, nongovernment organizations) who are active in the field of housing and urban

O. Serrat, *Anthropogenic Solutions for Climate Change*,
https://doi.org/10.1007/978-981-96-8654-4_7

development. The précis showcases some multilevel governance arrangements and pilot programs that Latin American and Caribbean cities have evolved, developed, and rolled out. Lastly, the précis passes comment on the operational, political, and financial feasibility of the recommendations it tables. In the immediate, the précis may inform the 28th yearly Conference of the Parties to the 1992 United Nations Framework Convention on Climate Change (COP28) that will gather global leaders, ministers, negotiators, climate activists, mayors and city managers, civil society representatives, and chief executive officers in Dubai from November 30 to December 12, 2023. The yearly COP conferences are intended for governments to agree on policies to limit global temperature increases and adapt to impacts from climate change.

7.2 Introduction

The 1990s were the heyday of globalization: the process was enshrined on December 11, 2001 when the People's Republic of China acceded to the World Trade Organization following 15 years of negotiations. Globalization tripled trade between 1990 and 2010 and had significant effects on economies, societies, and the natural environment (Vanham 2019). In the wake of the 2007–2008 financial crisis, a series of severe and mutually reinforcing global shocks (e.g., Brexit, China-U.S. trade tensions, the COVID-19 pandemic, the war in Ukraine) as well as growing concern for climate change turned the tide: terms such as "decoupling," "deglobalization," "localization," and "nearshoring" have entered the lexicon (Serrat 2023).

Globalization and population growth have fueled economic instability, social inequality, and environmental degradation worldwide (United Nations Department of Economic and Social Affairs 2021). The 2030 Agenda for Sustainable Development that the 193 United Nations Member States adopted on September 25, 2015 looks to reconcile people, planet, and profit, acting through 17 Sustainable Development Goals (SDGs) and 169 targets (United Nations General Assembly 2015). Relatedly, on December 12, 2015, the 196 members of the 21st yearly session of the Conference of the Parties to the 1992 United Nations Framework Convention on Climate Change (COP21), which aims to prevent dangerous anthropogenic interference with the climate system, signed the Paris Agreement on greenhouse gas emissions mitigation, adaptation, and finance. The Paris Agreement stipulates that, to limit global warming to 1.5°C, greenhouse gas emissions must peak before 2025 at the latest and decline 43% by 2030. Crucially, concrete actions are to be defined by each signatory to the Paris Agreement: and so, because commitments are nonbinding, the Paris Agreement lacks enforcement mechanisms (Conference of the Parties 2015). COP28, to be held in in Dubai from November 30 to December 12, 2023, will be a milestone moment when the world will take stock of its progress on the Paris Agreement.

Cities and their citizens can play a key role in achieving the 2030 Agenda and the Paris Agreement, even if neither was designed expressly for or by them, for instance by reducing their carbon emissions (Mukim and Roberts 2023; Organisation for Economic Co-operation and Development 2020; Vera et al. 2023). SDG 11 (Sustainable Cities & Communities), for instance, can synergize SDG 6 (Clean Water & Sanitation), SDG 7 (Affordable & Clean Energy), and SDG 13 (Climate Action), in the latter instance with global repercussions. Although cities occupy only 2% of Earth's land area, they are home to more than half of the world's population, consume 78% of the world's energy, and are responsible for more than 70% of global carbon dioxide emissions from final energy use (UN-Habitat 2023a; World Bank 2023). Energy drives lifestyles: globally, residential and commercial buildings consume more than half of all electricity (United Nations 2023a). Not surprisingly, the energy supply sector (i.e., electricity, heat, and other energy) is the largest contributor to global greenhouse gas emissions, responsible for 35% of the total (United Nations 2023b). By 2045, according to the World Bank (2023), the number of people living in cities will have swollen 1.5 times to 6 billion, adding 2 billion urban dwellers in toto; the world's total population will number 9.5 billion (from 8.1 billion today). For that reason, cities are tasked with delivering the bulk of the nationally determined contributions to the Paris Agreement that each country has committed to. Yet, there is more: on top of the challenges that owe to population growth, migration, and aging or inadequate infrastructure, especially in unplanned and informal settlements, high concentration of people and assets makes cities particularly vulnerable to the impacts of increasingly frequent climate-related risks (i.e., heat island effects from low albedo, heat-storing materials, urban canyons and urban geometry, waste heat, and lack of trees and other vegetation; urban flooding; and sea level rise) (Intergovernmental Panel on Climate Change 2022). And so, cities have no alternative but to engage with the 2030 Agenda and the Paris Agreement.

7.3 Problem Definition

The 2030 Agenda provides a shared blueprint for peace and prosperity for people and the planet, now and into the future (United Nations General Assembly 2015). The SDGs are not legally binding: nevertheless, countries are expected to take ownership and to establish a national framework for achieving them, with implementation and success relying on their own sustainable development policies, plans, and programs (United Nations General Assembly 2015). Countries are also responsible for follow-up and review at the national, regional, and global levels regarding the progress made in implementing the SDGs by 2030 (United Nations General Assembly 2015).

Conversely, the Paris Agreement is a legally binding international treaty on climate change that entered into force on November 4, 2016. The intent of the Paris

Agreement is to hold "the increase in the global average temperature to well below 2°C above pre-industrial levels" and pursue efforts "to limit the temperature increase to 1.5°C above pre-industrial levels" (Conference of the Parties 2015, p. 2).

The 2030 Agenda and the Paris Agreement apply to every country and region: both were endorsed by world leaders and both depend on action at local levels. The term "localizing" refers to the process of adapting something to a particular place or setting. In the context of the 2030 Agenda and the Paris Agreement, localizing relates to how local governments can offer support—framed by subnational contexts—through actions from the bottom up. Unsurprisingly, because they are permanent and highly organized centers of population, cities have core competencies in most of the policy areas that underlie the SDGs (e.g., climate change, housing, infrastructure, land use, transport, water) (Organisation for Economic Co-operation and Development 2020). What is more, city-scale actions lend themselves to a systems approach—that treats multiple infrastructure sectors as components of a larger working system—because cities are both the primary locus of demand and the level at which most infrastructure and services are provided (World Economic Forum 2022). Consequently, there is considerable value in popularizing knowledge of actions that have served to (a) strengthen capacity, (b) inform and raise awareness, (c) foster and reinforce cooperation, and (d) showcase innovative approaches that have boosted climate change mitigation and adaptation from the bottom up so city governments and actors of society elsewhere might take them up.

7.4 Geographical Focus

As of 2023, the population of Latin America and the Caribbean numbered almost 667 million, or about 8.3% of the total world population (Worldometer 2023). Latin America and the Caribbean is the most urbanized developing region on the planet, with 84% of the population living in cities such as São Paulo (22.6 million), Mexico City (22.3 million), Buenos Aires (15.5 million), Rio de Janeiro (13.7 million), Bogota (11.5 million), Lima (11.2 million), Santiago (6.9 million), Belo Horizonte (6.2 million), Guadalajara (5.4 million), and Monterrey (5.1 million) (Statista 2023). Comparing, the second most urbanized region in the world is North America (82.8%), followed by Europe (75.7%), Oceania (66%), Asia (52.6%), and Africa (44.7%) (Worldometer 2023). In line with the global average, about 70% of carbon dioxide emissions from Latin America and the Caribbean are generated in urban areas; furthermore, nearly 80% of climate-related disasters occur in urban areas (Villarroel and del Castillo 2023).

In an increasingly connected world, Latin America and the Caribbean was not able to thrive. Most of Latin America never globalized (or even internationalized) over the last 40 years because of weak governance, inequality, and insecurity; tellingly, intra-regional trade lagged even further with less than a fifth of trade (or 15%) taking place within the region itself (O'Neil 2022). Argentina and Brazil, for instance, remained two of the most closed economies in the world, with trade amounting to less than 30% of gross domestic product (O'Neil 2022). Latin America and the Caribbean regroups 33 countries, excluding dependencies and autonomous territories: as of 2021, based on gross national income (GNI) per capita, none were low-income countries (GNI per capita of up to $1085); five were lower-middle-income countries (GNI per capita of $1086 to $4255); 21 were upper-middle-income countries (GNI per capita of $4256 to $13,205); only six were high-income countries (GNI per capita of $13,206 or more); and one was not classified (Hamadeh et al. 2022). In a region exemplified by upper-middle-income economies, cities in Latin America and the Caribbean customized low-cost approaches to climate action, based on the characteristics of their territories and population, that may now be of interest to others cities in the world (Gallego Lizon 2022).

7.5 Question

Focusing on localized climate action in Latin America and the Caribbean, this précis underscores the benefits of a systems approach to the delivery and maintenance of sustainable urban infrastructure. In response to the problem definition, the question is: What approaches to localizing climate action might Latin American and Caribbean cities share with the world, especially lower-middle-income and upper-middle-income countries?

7.6 Methods

To anchor this précis and base it on evidence that solutions are at hand, the approach to data collection first involved a quick search of prominent publications using such keywords as *climate change mitigation*, *climate-smart cities*, *green infrastructure*, *low carbon cities*, *renewable energy in cities*, *sustainable urban development*, *urban climate adaptation*, *urban resilience*, and *urban sustainability*. The search showcased a preponderance of documents by the Inter-American Development Bank, the International Monetary Fund, the Organisation for Economic Co-operation and Development, and the World Bank. The same primary keywords were fed into the news tab of the Google search engine: that exercise showcased numerous contributions by the SDG Knowledge Hub (https://sdg.iisd.org/) in the International Institute for Sustainable Development, an independent think tank that promotes social leadership to create a world where people and the planet thrive; naturally, news agencies around the world report increasingly on related topics. Second, the aforementioned research question, accompanied by a description of the issue, was posted on ResearchGate to invite responses and raise public awareness. Third, the aforementioned research question, accompanied by the description of the issue, was shared by electronic mail to a subject matter expert in the Catholic University of Uruguay as well as with the Observatorio de la Energía y el Desarrollo Sustentable in that university (https://www.ucu.edu.uy/categoria/Observatorio-de-Energia-y-Desarrollo-Sus-401) on November 10 and November 15, 2023, respectively. Fourth, this on multiple occasions throughout November 2023, I interviewed the head of the housing and urban development division of a multilateral development bank. Fifth, courtesy of that same civil servant, I shared the aforementioned research question, accompanied by the description of the issue, with the participants to a regional dialogue on urban development held in Cartagena, Colombia on November 16–17, 2023; that roundtable, which involved 19 city mayors and city managers from eight countries of Latin America and the Caribbean, shared experiences of how subnational governments can lead action against climate change from the local level and elicited sundry recommendations and references. Knowledge gathered through data collection and wide-ranging feedback from interviews underpin the following issue analysis and corresponding proposed solutions.

7.7 Issue Analysis

Especially in Latin America and the Caribbean, cities house most of the global population and are epicenters of human activity. Cities can be termed "unsustainable" when they do not husband resources carefully to meet the needs of their inhabitants: cities that have acquitted the label include Kolkata (15.3 million), Jakarta (11.2 million), Lahore (13.9 million), and Rio de Janeiro (13.7 million) (World Population Review 2023). Other candidates to the list of unsustainable cities might include Delhi (32.9 million), Karachi (17.2 million), Mumbai (21.2 million), Mexico City (22.2 million), Nairobi (5.3 million), and São Paulo (22.6 million). Ten of the world's 15 most polluted cities by particulate matter concentration are in India (IQAir 2023). In Latin America, the 15 most polluted cities are in Chile and Peru (IQAir 2023).

To be sustainable, cities must be planned, designed, built (or reconfigured), and managed with consideration to their economic, environmental, and social impact (commonly referred to as the "triple bottom line") as well as resilience to climate change. To make cities and human settlements inclusive, safe, resilient, and sustainable in the face of increasing urbanization and climate change, what cities ought to aim for is comprehensively captured in Fig. 7.1, which lists the targets associated with SDG 11 (Sustainable Cities & Communities). Details of the indicators associated with the targets for SDG 11 (Sustainable Cities & Communities) are in the appendix.

Inevitably, the SDGs are top–down: nonetheless, this does not mean that local actors facing challenges must wait for national- and supranational-level actors to reach broadly-based and substantively effective agreements on what actions are called for. For sure, it took almost two decades of discussions, including five years of negotiations, for the world's governments to reach agreement on the new treaty to protect high seas marine life that was adopted on June 19, 2023; the treaty must now be ratified in national assemblies and then—somehow—acted upon. In the case of cities, however, local actors can—with local and practical vision—fully

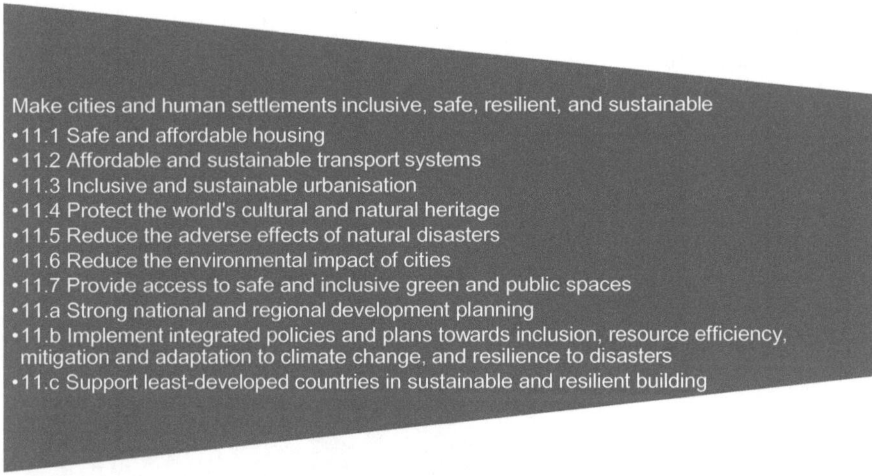

Fig. 7.1 SDG 11: Sustainable cities and communities—targets. (*Note.* From United Nations General Assembly (2015))

complement the 2030 Agenda and the Paris Agreement under their very specific conditions: it does not have to be the case that "Everybody talks about the weather, but nobody does anything about it," as both Mark Twain and Charles Dudley Warner are credited with saying.

In the case of cities, to repeat, action may be easier to take at local level than at national or supranational levels because high concentration of people and assets enables economies of scale and scope for higher relevance, coherence, effectiveness, efficiency, impact, and sustainability (Organisation for Economic Co-operation and Development 2023). Toward that, considering the targets associated with SDG 11 (Sustainable Cities & Communities), Figure 7.2 distinguishes with a systems approach the three sectors that from an environmental perspective condition much of life in cities and provide entry points for localizing climate action.

Looking to speed implementation of the SDGs in cities, Dulak (2023) highlighted the contribution of subnational authorities can make, for example by means of Voluntary Local Reviews (VLRs) that enable the review of SDG implementation at a local level. There is no standard structure for VLR documents; typically, however, they articulate (a) a statement by the mayor detailing the city's strategic vision, (b) the economic and social situation in the city, (c) an assessment of city performance using key performance indicators, highlighting progress towards achieving SDGs at the city level, (d) recommendations from the assessment of city performance, (e) the financial framework for supporting the city's efforts to achieve SDGs, and (f) conclusions and proposals for next steps (United Nations Economic Commission for Europe 2023). Helpfully, Deininger et al. (2019) had earlier designed a VLR handbook for cities wishing to take the lead on localizing the SDGs; without prescribing a one-size-fits-all approach, that handbook offers guidance to provide value to cities and their citizens. UN-Habitat (2023b) provided cutting-edge knowledge on VLRs and SDG localization in Latin American and

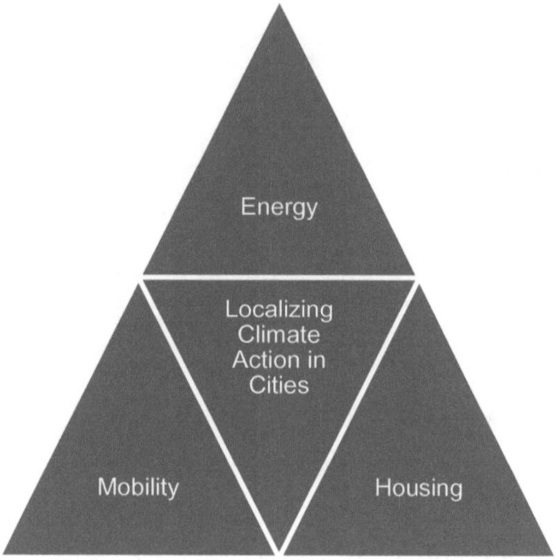

Fig. 7.2 Localizing climate action in cities with a systems approach

Caribbean cities from 2016 with recent updates from cities in Argentina, Bolivia, Brazil, Chile, Columbia, Costa Rica, Mexico, and Paraguay this last couple of years. To bolster global vision with examples of local and practical vision in Uruguay, Álvarez et al. (2022) presented different technologies and innovations that are enabling energy transition with information on costs, installed capacity and projections, as well as challenges.

Human-generated emissions have caused global warming hence climate change. The effects of climate change are hotter temperatures; more severe storms; increased drought; a warming, rising ocean; loss of species; not enough food; more health risks; and poverty and displacement (United Nations 2023b). Rising temperatures and heat waves are making cities less habitable, changes in precipitation patterns exacerbate water scarcity in urban areas, droughts and higher temperatures increase the risk of urban wildfires, and sea-level rise and storm surges pose threats to coastal cities (Talbot-Wright and Vogt-Schilb 2023).

7.8 Proposed Solutions

Global warming demands urgent action. As urbanization speeds up, cities must urgently decarbonize holistically (Gallego Lizon 2021). With more widespread use of clean energy, higher energy efficiency, increased and decentralized numbers of clean producers, and better demand management, cities must hone strategies in the areas of planning, designing, and building. In view of the targets associated with SDG 11 (Sustainable Cities & Communities) and cognizant of the intent of the Paris Agreement, Figure 7.3 frames across the energy—housing—mobility framework five recommendations to localize climate action in cities.

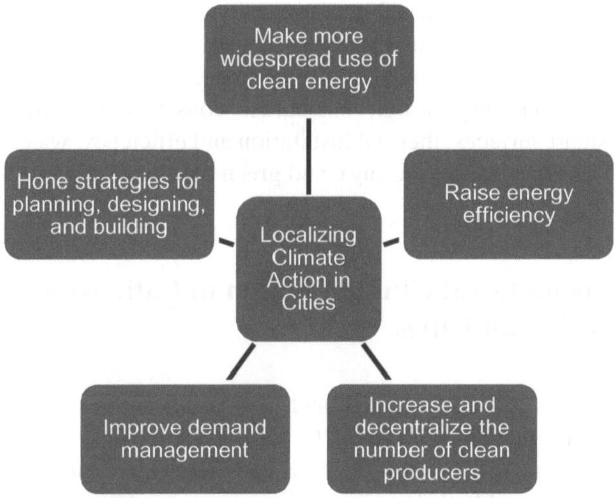

Fig. 7.3 Localizing climate action in cities

Specifically, with the help of VLRs, localizing climate action summons cities to:

- **Make more widespread use of clean energy.** The focal areas would relate to (a) electrification (e.g., electromobility, moving from biomass/liquefied petroleum gas to electricity), (b) waste-to-energy systems, and (c) geothermal/solar heating. The instruments would include investments and regulation.
- **Raise energy efficiency.** The focal areas would relate to (a) households (i.e., thermal, appliances, lighting), (b) power distribution, and (c) district heating/cooling. The instruments would include developer incentives. household subsidies, and standards.
- **Increase and decentralize the number of clean producers.** The focal areas would relate to (a) households (e.g., photovoltaic panels, small wind turbines), (b) municipalities (e.g., photovoltaic panels, hydropower). The instruments would include public–private partnerships (PPPs) and subsidies.
- **Improve demand management.** The focal areas would relate to (a) consumer behavior change (e.g., campaigning to produce behavioral change, implementing the sharing economy), (b) energy-saving tools for homes (e.g., smart sensors, energy-saving devices), and (c) system/voltage optimization. The instruments would include behavioral change campaigns, dynamic pricing, energy assessments, and policies and regulations.
- **Hone strategies for planning, designing, and building.** For the planning of cities, the focal areas would relate to (a) compact and mixed-use city planning and development, (b) reducing transit (e.g., treating the neighborhood as socioeconomic unit, raising efficiencies in transit-oriented development), and (c) efficient expansion. The instruments would include intelligent spatial plans and zoning. For the design and building of cities, the focal areas would relate to (a) bioclimatic design, (b) smart surfaces (i.e., reflective, permeable, carbon-

absorptive surfaces), (c) renewable construction materials (i.e., wood) and their reuse, and (d) biodiversity-positive approaches (i.e., nature-based solutions) and enabling carbon sequestration (i.e., green spaces) by design. The instruments would include building design standards, construction codes (e.g., flood protection, smart surfaces, thermal insultation and efficiency, water conservation and reuse, wind/hurricane-proofing), and green certification.

7.9 Showcasing Local Climate Action in Latin American and Caribbean Cities

Already, local actors in Latin America and the Caribbean have raised their ambitions for the decarbonization of cities. To set the scene, Case Study 7.1 outlines pell-mell a few recent, cross-cutting actions in Uruguay, a country that a short 15 years ago was using fossil fuels for a third of its energy generation. In 2007, Uruguay was using fossil fuels for a third of its energy generation: that country now runs on 98% renewable electricity thanks to biomass, solar, and wind parks. Next, this section points up initiatives that have had to do with clean energy; higher energy efficiency; increased and decentralized numbers of clean producers; better demand management; and new strategies for planning, designing, and building in Latin America and the Caribbean. Case Studies 7.1, 7.2, 7.3, 7.4, 7.5, and 7.6 illustrate effective local climate initiatives across Latin America and the Caribbean.

Case Study 7.1 Breaking "Green" news from Uruguay

- Uruguay banks on green energy.
  - City Level

    CUTCSA, Uruguay's largest urban public transport operator, has committed to replacing its entire fleet of buses with electrical vehicles.
  - National Level

    The government does not tax electric vehicles.

    The government plans to open a green hydrogen facility in Paysandu. Work will begin in 2024: it will require an investment of about $2 billion for green hydrogen and synthetic gasoline production and another $2 billion to install wind farm turbines that will power the plant.

    The Ministry of Economy and Finance has issued "green bonds" on the New York Stock Exchange; such bonds pay less interest if the country that emitted them accomplishes the goals to which it committed regarding climate change.

Note. Based on expert interview of November 10, 2023.

Case Study 7.2 Making more widespread use of clean energy

- Latin America's e-bus market is booming.
  - The number of e-buses in 32 Latin American cities is to increase from just over 4000 to more than 25,000 by 2030, with Brazil, Chile, Colombia, and Mexico expected to account for 82% of the region's e-bus market.
  - E-buses are expected to account for 30% of the 80,000 buses in the 32 Latin American cities by 2030, increasing to over 60% by 2050.
  - In Brazil, Chile, Colombia, and Mexico, the increase in electromobility owes to high urbanisation, high use of public transport, public policy that is focused on emissions reduction and sustainable urban mobility, and electric grid coverage with high penetration of renewable energies.

Note. Abridged from C40 Cities (2023).

Case Study 7.3 Raising energy efficiency

- Costa Rica, El Salvador, and Peru turn up the heat on geothermal energy.
  - Where available, geothermal energy can provide renewable power that is independent of weather conditions and not subject to international fuel price fluctuations. Geothermal energy can be used to, say, decarbonize the electricity grid, heat and cool buildings, and support industrial processes.
  - Costa Rica, El Salvador, and Peru have strong geothermal potential but differing levels of geothermal development.
  - Costa Rica looks to develop its remaining accessible geothermal capacity to export electricity to neighbors such as Nicaragua and Honduras, which rely on fossil fuels. Possibly also, Costa Rica may introduce geothermal heat pumps in the agrifood sector to replace conventional refrigeration powered by fossil fuels. Costa Rica's geothermal potential may also be developed for new uses, such as lithium mining or green hydrogen production.
  - El Salvador has for long used geothermal energy for power generation and its plants meet 25% of electricity consumption needs. Two geothermal fields are being developed through a build–operate–transfer model that attracts private investment by reducing the exploration risk and lessens the amount of capital the government needs to spend to develop the power plants.
  - Geothermal energy could be a boon to Peru, which produces 36% of its electricity from fossil fuels (but 58% from hydropower). Peru looks to encourage direct-use projects by establishing regulatory frameworks for geothermal heat.

Note. Abridged from Berg et al. (2023).

Case Study 7.4 Increasing and decentralizing the number of clean producers

- Mexico City reaps solar capital.
  - Mexico City's ambitious Solar City program was launched in 2019: it proposes to shift the capital to renewable energy.
  - As of 2021, more than 15,000 solar thermal systems had been installed in businesses, residential buildings, private households, and social housing units city-wide, driven by direct subsidies. As of 2021 also, about 1000 technicians had been trained to install solar systems, retrofit public buildings, promote waste-to-energy projects for biogas production, and increase the collection of used cooking oil to ramp up the production of biodiesel.

Note. Abridged from REN21 (2021).

Case Study 7.5 Improving demand management

- Latin America and the Caribbean fosters girl-led action on climate change.
  - Women and girls are disproportionately affected by climate change; however. they are for the most part excluded from the discourse on adaptation and mitigation.
  - Through the Youth and United Nations Global Alliance, the Food and Agriculture Organization of the United Nations has organized and led training workshops for 27 national girl guide association representatives from Argentina, Brazil, Chile, Dominica, St. Lucia, Mexico, Nicaragua, Panama, and Trinidad and Tobago. During the workshops, participants who will spearhead the Girl-Led Action on Climate Change program in their countries had the chance to learn more about climate change; reflect on how it is affecting their lives and those of the communities they live in; and appreciate why, as girls, they are impacted more than others. Most importantly, the girls realized how much depends on them as they discovered the many ways they can take climate action and make a difference. During the workshops, the participants co-created the educational curriculum for the climate change badge: specifically, they shared experiences, selected areas of highest interest, and included the topics that are most relevant to their local contexts.
  - The results that the Girl-Led Action on Climate Change program aims to achieve are: 12,000 girls have increased knowledge, skills, and support to take local action on climate change; 24,000 community members benefit from increased knowledge on climate change; 100 girls and young women are trained in girl-led advocacy; 3 young women repre-

sent the program by speaking at global policy events including the UNFCC Conference of the Parties; and 9 Girl Guiding/Scouting Associations have increased capacity to lead sustainable non-formal education programs focused on climate action.

Note. Abridged from FAO (2023).

Case Study 7.6 Honing strategies for planning, designing, and building

- Uruguay bets on wood to build sustainable social housing.
- Uruguay has established itself as the country of pulp: it has three processing plants and hundreds of thousands of hectares of eucalyptus, which are also its main export item. Now, Uruguay seeks to become the country of sustainable wooden homes with several projects in the offing.
- In the coming weeks, construction begins on a building that aims to become the tallest of its kind in South America. The material to be used, eucalyptus, emits less carbon and much as trees is capable of absorbing it.
- The seven-story construction, which is scheduled to begin before the end of 2023 in the city of Durazno, 180 kilometers north of Montevideo, will be destined for more than 50 social housing units. The building will be erected in less than a year.
- The project represents a milestone for the construction sector and for the future of the country: it promotes the use of wood as an alternative construction system for social housing.
- The wooden homes support the government's policy of reducing carbon emissions; they also strengthen local development and the circular economy.

Note. Abridged and translated from Cappelli (2023).

7.10 Limitations

The energy—housing—mobility framework advocated in this précis may well be universally applicable; on the other hand, the type, scale, cost, and pace of initiatives will often look different across cities. There can be no one-size-fits-all approach and capacity building of local governments will be needed: localizing climate action in cities hinges on experimental approaches to problem-solving and necessarily involves widespread and continuous stakeholder engagement, hence multilevel governance for deep collaboration among a diversity of social actors (Word Economic Forum 2022). The energy—housing—mobility framework and associated proposed solutions are operationally feasible and political feasibility will be enhanced by stakeholder engagement; then again, the financial burden of reducing climate change risks will be heavy for cash-strapped cities attempting to simultaneously modernize existing infrastructure and improve the quality of associated services.

7.11 Cost–Benefit Analysis

High concentration of people and assets in cities means that investments in localizing climate action will pay large dividends. Even so, project-based loans and grants from multilateral and bilateral development finance institutions and government budgets have to date been the primary instruments with which to disburse climate finance in low-income, lower-middle-income, and upper-middle-income countries. Finance to localize climate action should be scaled up. Necessarily, cities need to explore nontraditional approaches to project financing: these include municipal bonds, public–private partnerships, and energy savings performance contracts to pay for energy efficiency improvements with energy bill savings (Kelly et al. 2017). The investor base can also be diversified by attracting global banks, impact investing, institutional investors (e.g., insurance companies), investment funds, and philanthropic capital. Systemic change and investors will be encouraged by enabling policy and environments that remove impediments to transformation; here, policy-based finance and results-based finance can play valuable roles.

Appendix

SDG 11 Sustainable Cities and Communities: Targets and Indicators

Make cities and human settlements inclusive, safe, resilient, and sustainable

- 11.1 Safe and affordable housing
  - By 2030, ensure access for all to adequate, safe, and affordable housing and basic services and upgrade slums
- 11.2 Affordable and sustainable transport systems
  - By 2030, provide access to safe, affordable, accessible, and sustainable transport systems for all, improving road safety, notably by expanding public transport, with special attention to the needs of those in vulnerable situations, women, children, persons with disabilities, and older persons
- 11.3 Inclusive and sustainable urbanisation
  - By 2030, enhance inclusive and sustainable urbanization and capacity for participatory, integrated, and sustainable human settlement planning and management in all countries

- 11.4 Protect the world's cultural and natural heritage
  - Strengthen efforts to protect and safeguard the world's cultural and natural heritage
- 11.5 Reduce the adverse effects of natural disasters
  - By 2030, significantly reduce the number of deaths and the number of people affected and substantially decrease the direct economic losses relative to global gross domestic product caused by disasters, including water-related disasters, with a focus on protecting the poor and people in vulnerable situations
- 11.6 Reduce the environmental impact of cities
  - By 2030, reduce the adverse per capita environmental impact of cities, including by paying special attention to air quality and municipal and other waste management
- 11.7 Provide access to safe and inclusive green and public spaces
  - By 2030, provide universal access to safe, inclusive and accessible, green and public spaces, in particular for women and children, older persons, and persons with disabilities
- 11.a Support positive economic, social and environmental links between urban, peri-urban, and rural areas by strengthening national and regional development planning
- 11.b By 2020, substantially increase the number of cities and human settlements adopting and implementing integrated policies and plans towards inclusion, resource efficiency, mitigation and adaptation to climate change, resilience to disasters, and develop and implement, in line with the Sendai Framework for Disaster Risk Reduction 2015–2030, holistic disaster risk management at all levels
- 11.c Support least-developed countries, including through financial and technical assistance, in building sustainable and resilient buildings utilizing local materials

Note. From United Nations General Assembly (2015).

References

Álvarez A, Bastarrica F, Bremermann M, Di Chiara L, Estrada I, Ferrés F (2022) Monitor de innovación y nuevas tecnologías. Observatorio de la Energía y el Desarrollo Sustentable, Universidad Católica del Uruguay. https://carreras.ucu.edu.uy/aucdocumento.aspx?1478,2976

Berg R, Rubio J, Guzman A, Cohen K (2023) Turning up the heat on geothermal energy development in Latin America: varying scenarios for Costa Rica, El Salvador, and Peru. Center for Strategic and International Studies, March 10. https://www.csis.org/analysis/turning-heat-geothermal-energy-development-latin-america

C40 Cities (2023) Latin America's e-bus market is booming – new research shows where to invest, October 19. https://www.c40.org/news/latin-americas-e-bus-market/

Cappelli D (2023) Uruguay le apuesta a la madera para construir viviendas sociales sostenibles. El País, November 14. https://elpais.com/america-futura/2023-11-15/uruguay-le-apuesta-a-la-madera-para-construir-viviendas-sociales-sostenibles.html

Conference of the Parties (2015) Adoption of the Paris Agreement. FCCC/CP/2015/L.9/Rev/. https://unfccc.int/files/meetings/paris_nov_2015/application/pdf/paris_agreement_english_.pdf

Deininger N, Lu Y, Griess J, Santamaria R (2019) Cities taking the lead on the sustainable development goals: a voluntary local review handbook for cities. Carnegie Mellon University. https://www.brookings.edu/articles/a-handbook-to-help-cities-report-local-progress-on-the-sustainable-development-goals/

Dulak M (2023) Contribution of subnational authorities to multilateralism from the EU perspective—Implementation of the SDGs. Glob Policy 14(S.2):13–21

FAO (2023) Girl-led action on climate change in Latin America and the Caribbean. Youth and United Nations Global Alliance. https://www.fao.org/yunga/you/cc-lac/en/

Gallego Lizon T (2021) Raising the ambition of decarbonizing strategies for cities [PowerPoint slides]. Inter-American Development Bank.

Gallego Lizon T (2022) Weathering change: lessons for adaptive cities in Latin America and the Caribbean. In: Brears R (ed) The Palgrave encyclopedia of urban and regional futures. Palgrave Macmillan, pp 1–10

Hamadeh N, Van Rompaey C, Metreau E, Eapen M (2022) New World Bank country classifications by income level: 2022–2023. World Bank Blogs, July 1. https://blogs.worldbank.org/opendata/new-world-bank-country-classifications-income-level-2022-2023

Intergovernmental Panel on Climate Change (2022) Climate change 2022: impacts, adaptation, and vulnerability. Contribution of Working group II to the sixth assessment report of the Intergovernmental Panel on Climate Change. Cambridge University Press. https://www.ipcc.ch/report/ar6/wg2/

IQAir (2023) World's most polluted cities. https://www.iqair.com/world-most-polluted-cities

Kelly C, Martinez C, Hathaway-Williams W (2017). A framework for local action on climate change: 9 ways mayors can build resilient and just cities. American Progress, September 28. https://www.americanprogress.org/article/framework-local-action-climate-change/

Klaus I, Singer R (2018) The United Nations local authorities in four frameworks. Penn Institute for Urban Research

Mukim M, Roberts M (Eds) (2023) Thriving: making cities green, resilient, and inclusive in a changing climate. World Bank. https://openknowledge.worldbank.org/entities/publication/7d290fa9-da18-53b6-a1a4-be6f7421d937

O'Neil S (2022) Why Latin America lost at globalization—and how it can win now. Council on Foreign Relations, August 25. https://www.cfr.org/article/why-latin-america-lost-globalization-and-how-it-can-win-now

Organisation for Economic Co-operation and Development (2020) A territorial approach to the sustainable development goals: synthesis report. https://www.oecd.org/regional/a-territorial-approach-to-the-sustainable-development-goals-e86fa715-en.htm

Organisation for Economic Co-operation and Development (2023) Evaluation criteria. https://www.oecd.org/dac/evaluation/daccriteriaforevaluatingdevelopmentassistance.htm

REN21 (2021) How five cities are driving the energy transition in Latin America, June 3. https://www.ren21.net/five-cities-in-latin-america/

Serrat O (2023) Globalization and deglobalization: must it be either/or? Unpublished manuscript, Georgetown University.

Statista (2023) Most populated metropolitan areas in Latin America and the Caribbean in 2023. https://www.statista.com/statistics/1374285/largest-metropolitan-areas-in-latam/

Talbot-Wright H, Vogt-Schilb A (2023) Heat and high water: nine pathways to climate resilient development. Inter-American Development Bank. https://publications.iadb.org/en/heat-and-high-water-nine-pathways-climate-resilient-development
UN-Habitat (2023a) It's all about cities: we mustn't flip the coin on sustainable investment. https://unhabitat.org/news/14-nov-2023/its-all-about-cities-we-mustnt-flip-the-coin-on-sustainable-investment
UN-Habitat (2023b) Access cutting-edge knowledge on VLRs and SDG localization. https://unhabitat.org/topics/voluntary-local-reviews
United Nations (2023a) Act now: facts and figures. https://www.un.org/en/actnow/facts-and-figures
United Nations (2023b) Climate action: causes and effects of climate change. https://www.un.org/en/climatechange/science/causes-effects-climate-change
United Nations Department of Economic and Social Affairs (2021) Global population growth and sustainable development. UN DESA/POP/2021/TR/NO. 2. https://www.un.org/development/desa/pd/content/global-population-growth
United Nations Economic Commission for Europe (2023) Guidelines for the development of voluntary local reviews in the ECE region. ECE/HBP/2021/4. https://unece.org/info/Housing-and-Land-Management/pub/374943
United Nations General Assembly (2015) Transforming our world: the 2030 agenda for sustainable development. A/RES/70/1. https://sdgs.un.org/2030agenda
Vanham P (2019) A brief history of globalization. World Economic Forum, January 17. https://www.weforum.org/agenda/2019/01/how-globalization-4-0-fits-into-the-history-of-globalization/
Vera F, Uribe M, del Castillo S (2023) limate action and the Paris Agreement: the role of cities in Latin America and the Caribbean. Inter-American Development Bank. https://publications.iadb.org/en/climate-action-and-paris-agreement-role-cities-latin-america-and-caribbean
Villarroel K, del Castillo S (2023) Towards a sustainable future: climate action in Latin American and Caribbean cities. IDB Cities Network, September 19. https://blogs.iadb.org/ciudades-sostenibles/en/sustainable-future-climate-action-in-latin-american-caribbean-cities/
World Bank (2023) Urban development. https://www.worldbank.org/en/topic/urbandevelopment
World Economic Forum (2022) Delivering climate-resilient cities using a systems approach. https://www.weforum.org/publications/delivering-climate-resilient-cities-using-a-systems-approach/
World Population Review (2023) World cities. https://worldpopulationreview.com/world-cities
Worldometer (2023) Latin America and the Caribbean population. https://www.worldometers.info/world-population/latin-america-and-the-caribbean-population/

Part IV
Ecotopian Visions

Chapter 8
Bridging Crisis and Transformation: The *Half-Earth* Blueprint for Biodiversity's Revival

Abstract This précis considers Wilson's (Half-Earth: Our planet's fight for life. Liveright Publishing Corporation, 2016) *Half-Earth* proposal to address the biodiversity crisis. Inspired by innovative concepts from Schneider-Mayerson and Bellamy (An ecotopian lexicon. University of Minnesota Press, 2019) and themes from Butler (Parable of the sower. Four Walls Eight Windows, 1993), it identifies immediate actions, policy reforms, and shifts in behavior driven by grassroots activism, policy advocacy, and technological innovation. The précis also advocates for a Universal Declaration on Nature. Together, these elements provide a comprehensive blueprint for revitalizing biodiversity and fostering a sustainable coexistence with nature.

8.1 Half Empty or Half Full? Reimagining Our Relationship with Nature

Can you contemplate a future where half of Earth's surface remains untouched by human hands, where the web of life thrives, natural ecosystems flourish, and wildlife prospers once again? This is not an ecotopian fantasy. In *Half-Earth: Our Planet's Fight for Life* (2016), biologist and conservationist Edward O. Wilson presents a visionary plan to combat the ongoing mass extinction of animal and plant species.

In his book, Wilson identifies 35 "best places" in the biosphere across North America, the West Indies, South and Central America, Europe, Africa and Madagascar, Asia, Australia and Melanesia, Antarctica, and Polynesia, prioritized for conservation due to their extraordinary biodiversity richness or rarity and potential to contribute significantly to global biodiversity preservation. These areas must be kept inviolate. Wilson's idea is grounded in the principle that preserving vast, contiguous habitats is essential for maintaining biodiversity. By protecting large areas, we can ensure ecosystems remain intact and species have the space they need to flourish. This approach can prevent mass extinctions, mitigate climate change

O. Serrat, *Anthropogenic Solutions for Climate Change*,
https://doi.org/10.1007/978-981-96-8654-4_8

impacts, and secure a healthier, more resilient future for all living organisms, including us.

Wilson's radical yet achievable proposal offers hope, challenging us to reimagine our relationship with nature and take bold steps toward sustainability. Inspired by innovative concepts from *An Ecotopian Lexicon* (Schneider-Mayerson and Bellamy 2019) and transformative themes in the *Parable of the Sower* (Butler 1993), this précis explores the actions needed to link crisis with transformation, aiming to turn Wilson's visionary plan into reality.

8.2 In the Shadow of *The Sixth Extinction*

Throughout Earth's history, there have been five mass extinction events that drastically reduced biodiversity. These catastrophic events were triggered by natural phenomena such as volcanic eruptions, asteroid impacts, and climatic changes. The most recent and well-known of these is the Cretaceous-Paleogene extinction, which wiped out the dinosaurs around 66 million years ago.

Fast forward to the Anthropocene, we are witnessing another mass extinction event, the Holocene extinction, termed *The Sixth Extinction* by Kolbert (2014). This crisis, largely driven by human activities, has been exacerbated by the population surge that began in the nineteenth century. In 1800, the global population was 1 billion; by 1900, it had grown to 1.6 billion; and by 1960, it had almost doubled to 3 billion (Worldometer 2025). By 2020, the world population was approximately 7.8 billion (United Nations 2020). It is projected to reach 9.7 billion by 2050 (United Nations 2020). Although the global fertility rate has declined from 5.0 births per woman in 1950 to 2.5 births per woman in 2020, the absolute number of births remains substantial due to the large base population (World Bank 2020). Some regions, particularly Sub-Saharan Africa, still have high birth rates.

The rapid rise in human population fueled an escalating demand for resources, leading to widespread habitat destruction, overexploitation, pollution, climate change, and the inadvertent introduction of invasive species via global trade and travel. Consequently, the Holocene extinction profoundly affects a wide range of life forms, including mammals, birds, reptiles, amphibians, fish, and invertebrates. Alarmingly, species are disappearing at rates estimated to be 100–1000 times higher than the natural background rate (Pimm et al. 2014). Wildlife populations declined by an average of 69% between 1970 and 2018 (WWF 2022). Up to 1 million species are threatened with extinction, many within the coming decades (United Nations 2019). As of 2023, over 44,000 species are on the brink of extinction (IUCN 2023). Latin America has faced the steepest regional decline in average population abundance, plummeting by 94% between 1970 and 2018 (WWF 2022). Freshwater species populations have suffered the most substantial global decline, with an 83% reduction over the same period (WWF 2022). The consequences of this biodiversity loss extend far beyond the survival of individual species. It destabilizes entire ecosystems, jeopardizing the critical services they provide to humanity, such as pollination, water purification, and climate regulation (Cardinale et al. 2012).

8.3 Fluent in Life: Embracing a New Lexicon for Biodiversity Revival

Beginning with the idea that we lack the vocabulary to even describe environmental crises, Schneider-Mayerson and Bellamy (2019) introduce 30 loanwords that do not exist in English. Several of these loanwords are directly relevant when considering biodiversity extinction and the imperative of revival. For example, the concept of *apocalypso* can galvanize collective action by fostering community, creativity, and resilience. We can also unite communities in proactive environmental defense through *blockadia*, urging eco-exploitors and environmental degraders to "[r]espect existence, or expect resistance" (p. 38). We must strive for *dàtóng* or "great universality." Each one of us can become an *ildsjel*, dedicated to biodiversity revival. By recognizing and honoring *nahual*, our spiritual essence, we can promote compassionate and ethical treatment of animals and ecosystems. Reconnecting with *pachamama*, the vital energy that sustains all life, will foster reverence for the Earth. Embracing *plant time*—the slower, long-term cycles of plants—challenges short-term thinking and deepens our connection with nature's rhythms. By appreciating *qi*, the intrinsic energy that sustains all life forms and environments, we can promote ecological perspectives in the Anthropocene. *Sehnsucht* helps combine acceptance of past ecological harms with a hopeful vision for the future. A cultural attitude of *shikata ga nai*, also implying acceptance, can motivate decisive and unwavering efforts to confront ecological crises, recognizing that there is no alternative path forward. Appreciation of *sila*, which refers to such interconnected concepts as atmosphere, intelligence, spirit, and universe, can remind us that we are obliged to live in harmony with all life forms. *Shikata ga nai*: we must become more fluent in life.

8.4 Turning Point: A New Beginning

"As our task is new, so must we think anew," intoned Abraham Lincoln in his Annual Message to Congress on December 1, 1862, stressing the need for inventive solutions amid the Civil War. While some may view Wilson's (2016) audacious proposal as a non-starter in today's political climate, I find inspiration in that book's unspoken transformative themes. These themes resonate with those in Butler (1993), which tells the story of a young woman who creates a new belief system to survive and bring about societal change in a dystopian future marred by climate change, economic inequality, and social unrest. Some common themes in Butler and Wilson, that at heart have to do with crisis and transformation, are:

- **Destruction and Rebirth.** Butler's *Parable of the Sower* is set in a world experiencing societal collapse. Yet, the theme of rebirth and regeneration is central, underscoring the potential for new beginnings even amid destruction. Similarly, Wilson's proposal envisions reserving half of the Earth's land and seas for nature to regenerate.
- **Truth vs. Denial.** Lauren Olamina, Butler's protagonist, pursues truth and refuses to live in denial. This theme accentuates the necessity of confronting reality and working towards positive change. Wilson's proposal echoes this sentiment, advocating for proactive measures and policies to protect biodiversity and ensure the survival of ecosystems.
- **Acceptance of Change and Resiliency.** Lauren affirms the importance of accepting change and being resilient in the face of adversity. Her philosophy, "God is Change," pushes for proactive engagement. Likewise, Wilson's proposal requires a fundamental shift in how humanity interacts with the natural world, embracing change and resilience to ensure a sustainable future for all species.
- **Community and Empathy.** The transformative power of community and empathy is a recurring theme in Butler and Wilson. Lauren's Earthseed community thrives on cooperation, mutual support, and a shared vision for the future. Similarly, Wilson emphasizes the need for global cooperation and collective action, highlighting the importance of empathy and collaboration in conservation efforts.
- **Inclusion vs. Exclusion.** The *Parable of the Sower* explores the dynamics of gated communities and the exclusion of vulnerable populations. It critiques the societal tendency to isolate oneself from others, notably in times of crisis. In the same vein, Wilson's proposal calls for inclusive conservation efforts that involve all stakeholders, ensuring that marginalized communities and indigenous peoples are part of the solution and benefit from the preservation of natural habitats.
- **Religion, Hope, and Change.** Lauren creates her own belief system, Earthseed, which focuses on the idea that change is a divine force, suggesting that faith can be a source of strength and transformation. Wilson's vision also carries a sense of hope and a belief in humanity's ability to change its course.

The common themes in Butler and Wilson point up the importance of adaptability, community, and hope in tackling challenges and shaping a brighter future. By embracing change, fostering inclusivity, and cultivating empathy, we can work together to protect our planet and its diverse life forms. This collective effort embodies the spirit of rebirth and regeneration, ensuring a more hopeful future. Perhaps, Wilson's proposal will not sound so daunting if its proponents promulgate these messages with requisite ecotopian vocabulary.

8.5 Addressing the Biodiversity Crisis

Wilson's (2016) *Half-Earth* proposal bridges crisis and transformation, urging us to dauntlessly reimagine our relationship with nature by designating more protected areas. Many of these areas are already protected or partially protected; for example, Serrat (2024) elaborates on the Heart of Borneo Initiative, which exemplifies this effort. However, the *Half-Earth* project is dedicated to expanding and connecting them for comprehensive conservation. While concerns about feasibility persist, the main worry is the potential displacement of indigenous and local communities, even if Wilson's proposal does not call for this. Indigenous and local communities have not caused biodiversity loss; the root causes lie in unsustainable consumption, economic inequality, and environmental degradation in other regions. Indeed, these communities may welcome relief from illegal loggers, poachers, and unauthorized settlers.

"Never underestimate the power of an idea whose time has come," Victor Hugo is alleged to have said. Acting on Wilson's proposal requires a multifaceted and creative approach to redesigning the world, including immediate actions, policy reforms, and shifts in societal behavior:

- **Immediate Actions**
  - **Protected Areas.** Expanding and strengthening protected areas, such as national parks and wildlife reserves, is essential for preserving biodiversity. These areas provide safe havens for endangered species and help maintain ecosystem functions.
  - **Restoration Projects.** Implementing restoration projects to rehabilitate degraded ecosystems can enhance biodiversity and improve ecosystem services. This includes reforestation, wetland restoration, and the removal of invasive species.
- **Policy Reforms**
  - **Economic Incentives.** Providing economic incentives for conservation efforts, such as subsidies for sustainable land use practices and penalties for habitat destruction, can drive positive change.
  - **International Cooperation.** Establishing a global consensus on the importance of biodiversity conservation is crucial. Countries must work together to create policies that prioritize the protection of natural habitats and the preservation of species.
- **Shifts in Behavior**
  - **Community Engagement.** Engaging local communities in conservation efforts can empower them to take ownership of their natural resources. Community-led initiatives, such as reforestation projects and wildlife protection programs, can have a significant impact.

- **Environmental Ethics.** Promoting an ethic of environmental stewardship can shift societal values towards greater respect for nature. Education programs that underscore the interconnectedness of all living beings and the importance of biodiversity can foster a sense of responsibility towards the environment.

The methods of change needed to spur immediate actions, policy reforms, and shifts in behavior include grassroots activism, policy advocacy, and technological innovation.

- **Grassroots Activism.** Community-led initiatives play a pivotal role in raising awareness and driving change. Grassroots organizations can mobilize resources, advocate for policy changes, and provide support to those affected by biodiversity loss.
- **Policy Advocacy.** Lobbying for legislative changes that prioritize biodiversity conservation and sustainable practices is essential. Engaging with policymakers, participating in international forums, and supporting organizations dedicated to environmental protection can amplify these efforts.
- **Technological Innovation.** Leveraging technology to develop sustainable solutions, such as renewable energy sources and eco-friendly materials, we can enhance environmental resilience and reduce the impact of human activities on ecosystems.

8.6 Time's up: Envisioning a Universal Declaration on Nature

In the age of the Anthropocene, nature needs an equivalent to the Universal Declaration of Human Rights (United Nations 1948). I see this conclusion as a natural progression from pivotal historical documents: the Stockholm Declaration (1972), the World Charter for Nature (1982), the Rio Declaration on Environment and Development (1992), the Convention on Biological Diversity (1992), Agenda 21 (1992), the Aarhus Convention (1998), the Earth Charter (2000), and the Paris Agreement (2015) (Serrat 2012). To my mind, the core principles of a Universal Declaration on Nature, as outlined in Fig. 8.1, would likely include biodiversity conservation, equity and justice, global cooperation, interdependence, intrinsic value, respect and stewardship, and sustainable use.

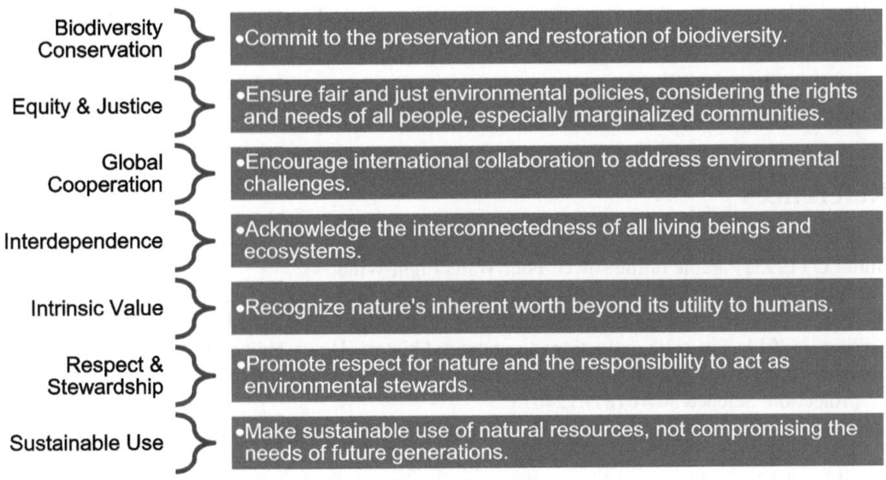

Fig. 8.1 Core principles of a universal declaration on nature

8.7 Every End, a Beginning: Bridging Crisis and Transformation

The vision presented by Wilson (2016), energized by the fresh thinking of Schneider-Mayerson and Bellamy (2019) and the transformative themes in Butler (1993), offers a compelling blueprint for revitalizing biodiversity. We can take immediate actions, enact meaningful policy reforms, and reshape societal behavior to foster a more resilient and harmonious coexistence with nature. As a critical first step, establishing a Universal Declaration on Nature would provide the rationale to bridge the current crisis with a transformative future. *Sila*!

8.8 Postscript

There are concerns that Wilson's (2016) *Half-Earth* proposal could reinforce the notion that humans and nature are inherently incompatible. Could the areas outside the designated preserves become ecological wastelands, serving as the antithesis of biodiverse zones? Moreover, would Wilson's vision inspire individuals to adopt more sustainable behaviors, or simply lead them to withdraw from the protected areas? Proponents argue that the proposal is not about isolating humans from nature but about creating spaces where ecosystems can flourish without human disturbance. This approach has the potential to foster a broader cultural shift, encouraging society to value and integrate biodiversity into human environments. However, the success of this proposal would heavily depend on its implementation and messaging. If the preserves are perceived as "off-limits," public disengagement could follow. Conversely, framing these zones as a shared responsibility, coupled with education and incentives, might promote sustainable practices and deepen humanity's connection to nature.

References

Butler O (1993) Parable of the sower. Four Walls Eight Windows

Cardinale B et al (2012) Biodiversity loss and its impact on humanity. Nature 486(7401):59–67

IUCN (2023) The IUCN red list of threatened species. https://www.iucnredlist.org

Kolbert E (2014) The sixth extinction: an unnatural history. Henry Holt and Company

Pimm S et al (2014) The biodiversity of species and their rates of extinction, distribution, and protection. Science 344(6187):1246752

Schneider-Mayerson M, Bellamy B (eds) (2019) An ecotopian lexicon. University of Minnesota Press

Serrat O (2012) World sustainable development timeline. Asian Development Bank. https://www.adb.org/publications/world-sustainable-development-timeline

Serrat O (2024) Preserving paradise: the heart of Borneo Initiative. Unpublished manuscript, Georgetown University. https://www.researchgate.net/publication/381993897_Preserving_Paradise_The_Heart_of_Borneo_Initiative

United Nations (1948) Universal Declaration of Human Rights. https://www.un.org/en/about-us/universal-declaration-of-human-rights

United Nations (2019) Global assessment report on biodiversity and ecosystem services. https://www.un.org/sustainabledevelopment/biodiversity/

United Nations (2020) World population prospects 2020. Department of Economic and Social Affairs, Population Division. https://population.un.org/wpp/

Wilson E (2016) Half-Earth: our planet's fight for life. Liveright Publishing Corporation

World Bank (2020) World development indicators 2020. https://data.worldbank.org/

Worldometer (2025) World population by year. https://www.worldometers.info/world-population/world-population-by-year/

WWF (2022) Living planet report 2022. https://www.worldwildlife.org/publications/living-planet-report-2022

Chapter 9
Building Sustainable Futures: Key Themes in Ecotopian Visions

Abstract This précis delves into the concept of ecotopia, a term coined to describe a society where humanity and nature coexist in harmony. It explores three interlinked themes—contrasting worlds, interconnectedness and liberation, and affective dimensions—to illuminate the challenges of contemporary environmental issues and offer valuable insights into creating a more sustainable and harmonious future. Dystopias are imagined societies where everything has gone wrong, often characterized by environmental catastrophes, oppressive governments, pandemics, resource depletion, social unrest, technological failures, or other forms of societal collapse or strife. Notable examples include Huxley (1932), Orwell (1949), and Collins (2008). These dystopias serve as a stark contrast to utopias, which depict idealized, perfect societies. Ultimately, utopias and dystopias are compelling because they reflect human potential exaggerated on a grand scale.

9.1 Ecotopian Visions: Harmonizing Humanity and Nature

Imagine a world where humanity and the natural environment coexist in harmony—a true *ecotopia*. This term, coined by Callenbach (1975), explores the challenges and possibilities of creating a just and sustainable society. Although Callenbach introduced the term, the concept of a harmonious balance between humans and nature has been a part of human imagination and philosophy long before his work. *The I Ching* (1046–771 BCE), Hesiod's *The Works and Days* (eighth century BCE), Laozi's *Tao Te Ching* (sixth century BCE), *The Bhagavad Gita* (second century BCE), *The Kojiki* (712 CE), as well as more recent texts like Thoreau (1854), Leopold (1949), Carson (1962), and Seuss (1971), are prime examples. In this précis, drawing from both fictional and nonfictional literary foundations, I consider three interlinked themes that provide a comprehensive framework for understanding the intricate interplay of psychological and sociocultural factors in ecotopian visions: contrasting worlds, interconnectedness and liberation, and affective dimensions. These themes are discussed in this order to progressively build an understanding of the multifaceted nature of ecotopian visions, from external societal structures

O. Serrat, *Anthropogenic Solutions for Climate Change*,
https://doi.org/10.1007/978-981-96-8654-4_9

to deeply personal emotional and psychological experiences. They illuminate the challenges we face in addressing contemporary environmental issues and offer valuable insights into the possibilities for creating a more harmonious and sustainable future.

9.2 Contrasting Worlds

A central theme in ecotopian visions is the exploration of contrasting worlds, exemplified by Le Guin's (1974) *The Dispossessed.* Written around the same time as Callenbach (1975), Le Guin also reflects the growing environmental consciousness and desire for social change of the 1960s and 1970s, a movement significantly influenced by Carson (1962). Le Guin presents a dual narrative of two contrasting societies: the anarchist society of Anarres and the capitalist society of Urras. This juxtaposition highlights the tension between utopian ideals and the practical challenges of implementing them. Anarres represents a society striving for equality and communal living, but it suffers from bureaucratic rigidity, scarcity, and social conformity. Urras embodies the inequalities and excesses of capitalism, leading to social stratification, environmental degradation, and exploitation. Through this contrast, Le Guin invites readers to reflect on the feasibility of utopian ideals and the compromises required to achieve them (Serrat 2025).

Singh (2022) adds another layer to the theme of contrasting worlds by framing alternative utopias that are neither purely idealistic nor dystopian but present practical and sustainable alternatives. Informed by responses to colonialism, Singh sees three kinds of utopias: exclusionary/reactionary; mimetic/modernist; and local/planetary. Based on Singh's compelling argument, "[l]ocalized utopias (of the third kind) are perhaps the only real utopias possible, because geography, culture, history, and […] manifestations […] shape, in part, how people conceive and build their ideals of the best life" (Singh 2022, p. 29). Without romanticizing the local, Singh's work inspires us to envision and create new, more sustainable worlds right in our own backyards, rather than on distant planets. These localized utopias address the shortcomings of both utopian and dystopian visions, whether grand or grotesque, by suggesting practical and achievable alternatives for a better future. The Global Tapestry of Alternative (n.d.) (https://globaltapestryofalternatives.org/) showcases real-world, transformative alternatives that create paths towards cultural and knowledge diversity, economic and political democracy, ecological resilience, localized self-reliance, and social justice.

The exploration of contrasting worlds also extends to the examination of different economic systems. In Le Guin (1974), the author delves into the economic structures of Anarres and Urras, highlighting the stark differences between the anarchist and capitalist systems. Anarres operates on principles of mutual aid and communal ownership, while Urras is characterized by private property and economic inequality. This contrast prompts readers to question the ethical implications and sustainability of different economic models. Le Guin ultimately suggests that, although neither system is perfect, the values of communal ownership and mutual aid offer a more equitable and sustainable path forward.

Gender roles are another crucial aspect of contrasting worlds. It is crucial to address this topic, as women constitute roughly 50% of the global population. In Le Guin (1974), the author's portrayal of gender equality on Anarres challenges traditional gender norms and highlights the potential for more egalitarian societies. Women on Anarres participate equally in all aspects of society, including leadership roles, manual labor, and scientific research, reflecting a commitment to gender equality. In contrast, Urras mirrors the patriarchal structures prevalent in many contemporary societies. Women on Urras are relegated to subordinate roles, with limited access to education and professional opportunities, and are expected to adhere to traditional domestic responsibilities. By presenting these contrasting worlds, Le Guin encourages readers to envision alternative social structures that promote gender equality and social justice.

"In the beginning was the Word, and the Word was with God, and the Word was God" (John 1:1, Holy Bible 2011). Put simply, words shape our understanding of the world and our place within it. They are essential for thinking, expressing thoughts, communicating complex ideas, and guiding actions. Schneider-Mayerson and Bellamy (2019) emerged as a response to the growing need for a specialized vocabulary to articulate and address the complex environmental challenges of our time. By providing precise and evocative terminology, Schneider-Mayerson and Bellamy help us think critically about our relationship with the natural world and foster a deeper understanding of ecological concepts. Schneider-Mayerson and Bellamy introduce the terms *apocalypso* and *blockadia* to illustrate the theme of contrasting worlds. *Apocalypso* merges "apocalypse" and "calypso," evoking a dance with disaster and capturing cultural and emotional responses to environmental crises. *Blockadia* refers to the global resistance movement against fossil fuel extraction and environmental destruction. These terms emphasize the cultural narratives and myths that influence public perception and reactions to environmental issues, underscoring the significance of scrutinizing the stories we create and share about the environment.

9.3 Interconnectedness and Liberation

Another central theme in ecotopian visions is that of interconnectedness and, in seeming paradox, liberation. Schneider-Mayerson and Bellamy (2019) encapsulate this theme through its entries *heyiya* and *total liberation*. *Heyiya*, a term from the Lakota language spoken by a Native American tribe primarily located in the northern Great Plains, symbolizes the interconnectedness of all life and the cyclical nature of existence. It emphasizes the importance of recognizing and respecting the interdependence of all living creatures and the environment. By incorporating *heyiya* into our vocabulary, we can foster a deeper understanding of ecological relationships and promote a more holistic approach to environmental stewardship. Relatedly, *total liberation* advocates for the interconnected liberation of humans, animals, and the environment. It emphasizes that true freedom and justice cannot be achieved unless all forms of oppression and exploitation are addressed simultaneously. *Total liberation* encourages a holistic approach to activism, recognizing that the well-being of all living creatures and the planet are deeply intertwined. By embracing *heyiya* and *total liberation*, we can work towards a more just and sustainable world.

Recognizing that, ever since Plato wrote *Republic* (ca. 380 BCE), utopias have often been visualized in terms of the city, Mumford (1965) delves deeper into the relationship between urban planning, technology, and the pursuit of utopian ideals. Mumford highlights the authoritarian aspects of many utopias, suggesting that the structured and controlled nature of cities often leads to such characteristics. He asks, "How could the human imagination, supposedly liberated from the constraints of actual life, be so impoverished?" (Mumford 1965, p. 278). He examines the interplay between technological advancements and social structures in utopian visions, emphasizing the need for a balance between technological progress and human values. Mumford critiques the positivist approach that overlooks human capacities for potential, purpose, and ideals, and advocates for a utopian vision that respects these human capacities and serves as a corrective to the limitations of positivism. The design of our cities and the technologies we use play a crucial role in shaping our societies and their environmental impact, he presciently intuits. And so, by reimagining urban spaces and technological systems, we can create more sustainable and equitable communities that reflect the principles of interconnectedness and liberation.

9.4 Affective Dimensions

In Weik von Mossner (2017), the author leverages ecocriticism to explore how empathy and emotional engagement with the environment can foster a deeper connection and drive for ecological preservation. Weik von Mossner reasons that our minds are both embodied in a physical body and embedded in a physical environment. By engaging with environmental narratives that evoke strong emotional responses, Weik von Mossner affirms that we can cultivate a sense of empathy and urgency that motivates action. This approach underscores the importance of addressing the emotional and psychological dimensions of climate change to create lasting and meaningful change. However, the exploration of emotional and psychological dimensions can also reveal cognitive dissonance and denial. Referencing Norgaard (2011) on the painful emotions associated with climate change, Weik von Mossner lists three different forms of denial as coping mechanisms: literal, interpretive, and implicatory. Literal denial leads people to outright reject the reality of climate change. Interpretive denial involves acknowledging the facts but downplaying their significance or interpreting them in a way that minimizes the threat. Implicatory denial occurs when individuals accept the reality and significance of climate change but fail to act on it, often rationalizing their inaction. These psychological mechanisms help explain why people may fail to act on environmental issues despite being aware of the problem (Norgaard 2011; Weik von Mossner 2017).

The role of media representation in shaping public perception and response to climate change is another important aspect of this theme (Weik von Mossner 2017). Evidently, the way climate change is portrayed in the media (e.g., audiobooks, blogs, films, graphic narratives, interactive exhibits, literary texts, music, news articles, podcasts, social media, stage performances, television formats, video games, etc.) can influence how people perceive the issue and their sense of urgency. Conspiracy theories, disinformation, fake news, hoaxes, misinformation, propaganda, and rumors can create confusion with conflicting messages and diminish the motivation to act. By carefully examining environmental narratives across media, Weik von Mossner (2017) assures us, we can better understand the barriers to effective communication and develop strategies to improve public awareness and engagement.

Schneider-Mayerson and Bellamy (2019) provide insight into the emotional and psychological dimensions of climate change. *Sila*, a term from Inuit culture, represents the weather, climate, and the breath of life, emphasizing the interconnectedness of natural elements. It reminds us of the deep emotional and spiritual connection that many indigenous cultures have with the environment. By understanding and respecting these connections, we can foster a greater sense of empathy and responsibility towards the natural world. *Solastalgia*, a portmanteau word, describes the emotional distress caused by environmental change, akin to feeling homesick while still at home. This term captures the profound sense of loss and displacement that many people experience because of climate change. By acknowledging and addressing these emotional responses, we can better understand the barriers to action and develop more effective strategies for engaging people in environmental stewardship.

9.5 Integrating Ecotopian Themes for a Sustainable Future

In ecotopian visions, the recurring theme of contrasting worlds highlights the tension between utopian ideals and practical realities, encouraging us to thoughtfully examine the stories we tell ourselves about the environment. The likewise recurring theme of interconnectedness and liberation emphasizes the importance of holistic approaches to activism and environmental stewardship, recognizing the deep interdependence of all living creatures and the planet. And, the frequent theme of affective dimensions underscores the need to address the emotional and psychological barriers to action, fostering empathy and emotional engagement with the environment. By discerning these themes in ecotopian visions, we can develop more effective and comprehensive strategies for addressing the environmental challenges of our time. Ultimately, it is through this integrated understanding that we can work towards a more just, sustainable, and ecologically harmonious world.

The exploration of contrasting worlds, interconnectedness and liberation, and affective dimensions also highlights the importance of interdisciplinary approaches to environmental issues. By drawing on insights from environmental science, literature, psychology, and sociology we can develop a more nuanced understanding of the factors that influence our responses to climate change. This interdisciplinary perspective is essential for developing effective strategies to address the complex and multifaceted nature of environmental challenges.

Furthermore, the examination of ecotopian visions encourages us to think creatively and imaginatively about the future. By envisioning alternative worlds and exploring the possibilities of sustainable living, we can inspire hope and motivation for positive change. This imaginative approach is crucial for overcoming the sense of despair and helplessness that often accompanies discussions of climate change, and for developing effective and comprehensive strategies to address the environmental challenges of our time and work towards a more just, sustainable, and ecologically harmonious world.

References

Callenbach E (1975) Ecotopia: the notebooks and reports of William Weston. Banyan Tree Books
Carson R (1962) Silent spring. Houghton Mifflin
Collins S (2008) The hunger games. Scholastic Press
Global Tapestry of Alternatives (n.d.) Global tapestry of alternatives. https://globaltapestryofalternatives.org/
Holy Bible (2011) New international version. Biblica, Inc.
Huxley A (1932) Brave new world. Chatto & Windus
Le Guin U (1974) The dispossessed. Harper & Row
Leopold A (1949) A sand county almanac. Oxford University Press
Mumford L (1965) Utopia, the city, and the machine. Daedalus 94(2):271–292
Norgaard K (2011) Living in denial: climate change, emotions, and everyday life. MIT Press
Orwell G (1949) 1984. Secker & Warburg
Schneider-Mayerson M, Bellamy B (eds) (2019) An ecotopian lexicon. University of Minnesota Press
Serrat O (2025) Exploring ecotopian visions in Le Guin (1974). Unpublished manuscript, Georgetown University. https://www.researchgate.net/publication/388653278_Exploring_Ecotopian_Visions_in_Le_Guin_1974
Seuss D (1971) The Lorax. Random House
Singh V (2022) Utopias of the third kind. PM Press
Weik von Mossner A (2017) Affective ecologies: empathy, emotion, and environmental narrative. Ohio State University Press

Chapter 10
Exploring Ecotopian Visions in Le Guin (1974)

Abstract This précis examines ecotopian ideals in Le Guin's (The dispossessed. Harper & Row, 1974) novel. It explores concepts such as communal ownership, decentralized governance, and sustainable living, highlighting their relevance to contemporary issues. The précis considers Le Guin's cautionary perspective, echoing the concerns raised by Mumford (Utopia, the city, and the machine. Daedalus 94(2):271–292, 1965) regarding the technical and practical challenges of realizing such ideals. Additionally, it incorporates Singh's (Utopias of the third kind. PM Press, 2022) concept of "utopias of the third kind," advocating for adaptable and realistic visions of the future.
Le Guin's (1974) *The Dispossessed* is a profound examination of ecotopian ideals depicting a world, Anarres, that prioritizes ecological balance, social equality, and communal values. This précis delves into three ecotopian concepts from the novel: communal ownership, decentralized governance, and sustainable living. These concepts are highly relevant to the pressing ecological, economic, and societal issues our planet faces today. However, the précis acknowledges that Le Guin's perspective on these and other ecotopian themes is cautionary, echoing the problematic technicalities highlighted by Mumford (1965). While ecotopias offer tantalizing visions of harmonious futures, they must be more firmly rooted in the realities of our world to truly thrive. In conclusion, the précis invites consideration of "utopias of the third kind" as defined by Singh (2022).

10.1 On Brave New Worlds

Imagine a society where the very essence of life is intertwined with ecological balance, communal values, and sustainable living (Serrat 2025). Le Guin (1974) invites us into such a world. However, there is a catch: Le Guin subtitled her novel *An Ambiguous Utopia* to remind us that ecotopias may not be what they seem (or are advertised to be). By dialectically juxtaposing Anarres, an anarcho-communist world, with Urras, a capitalist society on another planet, Le Guin challenges us to

O. Serrat, *Anthropogenic Solutions for Climate Change*,
https://doi.org/10.1007/978-981-96-8654-4_10

reflect on ecological, economic, political, psychological, and social themes through the sensibilities and experiences of Shevek, the book's protagonist.

10.2 Exploring Anarres: A Deep Dive into Ecotopian Ideals

Five ecotopian themes in Le Guin (1974) resonated with me:

- **Collectivism vs. Individualism.** Anarres values collectivism and operates on anarcho-communist principles, emphasizing communal ownership, equality, and mutual aid. Conversely, Urras prioritizes individualism and competition within a capitalist society: wealth and power are concentrated in the hands of a few, engendering economic disparities at the expense of social cohesion.
- **Gender Equality.** Gender equality is ingrained in the social fabric of Anarres, where men and women share responsibilities. In contrast, Urras is characterized by patriarchy: women face discrimination and must conform to specific roles, limiting their opportunities and freedoms.
- **Decentralization vs. Hierarchical Power Structures.** Anarres strives for a decentralized society without hierarchical power structures, promoting equality and shared decision-making. On the other hand, Urras is characterized by rigid hierarchies and centralized authority, perpetuating social inequality.
- **Spiritual Fulfillment vs. Material Wealth.** On Anarres, material wealth is de-emphasized in favor of spiritual fulfillment. Then again, in Urras, material wealth and consumerism are central to the society, often overshadowing deeper values.
- **Sustainability vs. Environmental Exploitation.** Anarres practices sustainable living, with a focus on resource conservation and minimalism. Quite the opposite, Urras engages in environmental exploitation for economic gain, resulting in ecological degradation.

10.3 Beyond the Facade: Utopias Unveiled

Even though Anarres initially appears to be an ecotopia, the truth is that both Anarres and Urras are plagued by systemic problems. By examining the five ecotopian themes mentioned above through my own experiences and viewpoints, I can identify three characteristics of Anarres—community ownership, decentralized governance, and sustainable living—that, due to their idealization, rigid implementation, and lack of integration, ultimately impede the advancement of humanistic values and technological progress, prompting Shevek to question the true nature of freedom. In each instance, I contrast these with the realities of Urras to yield a clear-eyed understanding of the complexities and contradictions within both societies.

- **Communal Ownership.** Anarres mandates communal ownership to promote equality, reduce economic disparities, and foster a sense of community and mutual support. However, this system has led to inefficiencies, lack of personal incentives, and potential for resource mismanagement. Shevek experiences frustration with bureaucratic inefficiencies and the lack of personal motivation among some individuals. In stark contrast, Urras encourages individual ownership and personal responsibility to foster economic growth. Yet, this approach has resulted in wealth disparities, social inequality, and potential exploitation of resources and people. The stark contrast between the wealthy elite and the impoverished masses on Urras is evident in Shevek's observations during his visit.

- **Decentralized Governance.** Anarres operates on anarcho-communist principles, emphasizing local decision-making, communal ownership, and equality. This decentralized approach has encouraged community involvement and reduced the concentration of power. However, it has also led to fragmented policies, lack of coordination, and potential for local conflicts. Shevek's struggles with the Syndicate of Initiative highlight the challenges of decentralized governance. Conversely, Urras has a more centralized governance structure, providing clear direction and efficient decision-making. However, this centralization has led to authoritarianism, concentration of power, and suppression of dissent. The oppressive government of A-Io on Urras exemplifies the dangers of centralized power.
- **Sustainable Living.** Anarres promotes environmental conservation, reduces ecological footprint, and encourages a minimalist lifestyle. However, this focus on sustainability limits economic growth, technological advancement, and personal comforts. The harsh living conditions and scarcity of resources on Anarres are a constant challenge for its inhabitants. On the other hand, Urras prioritizes economic growth and technological advancements, which, at least in theory, can lead to improved living standards and innovation. However, this has also resulted in environmental degradation, resource depletion, and unsustainable practices. Shevek is struck by the environmental exploitation and pollution he witnesses on Urras.

10.4 Anarres and Urras Through Mumford's (1965) Lens

Mumford (1965) highlights the potential dangers of unchecked technological progress and the need for societies to prioritize ecological and social well-being over mere economic growth. Applying Mumford's perspective to Le Guin (1974), we can see a clear alignment with his cautionary stance. Le Guin's depiction of Anarres, with its focus on anarcho-communist principles, echoes Mumford's advocacy for a more human-centered approach to societal development. However, Le Guin also acknowledges the challenges and limitations of such a society, much like Mumford does. Le Guin presents Anarres as an ambiguous utopia where systemic problems and inefficiencies persist despite the noble ideals, leaving its ultimate value and feasibility open to interpretation. In contrast, Urras represents the pitfalls of a society driven by material wealth and technological advancement without regard for social and ecological balance. This dichotomy between Anarres and Urras mirrors Mumford's critique of modern industrial societies and his call for a more balanced and sustainable approach to progress. By examining Le Guin through Mumford's lens, we gain a nuanced understanding of the complexities and contradictions within both societies, and the importance of striving for a harmonious balance between humanistic values and technological progress.

10.5 Utopias of the Third Kind

Singh (2022) argues that utopias should be adaptable, resilient, and rooted in the realities of our current world, rather than being purely idealistic or unattainable visions. Applying this perspective to Le Guin (1974), much as we did using Mumford's (1965) lens, we can see that Anarres does not align with Singh's idea of a "utopia of the third kind."

> Utopias of the third kind [...] are visions that are grounded in the local, in its geography and social–cultural–ecological surround, but locate themselves in a planetary context where "local" and "planetary" are not mutually exclusive categories but are connected in space and time. (Singh 2022, p. 27)

Several improvements might be made to transform Anarres into a more effective utopia as envisioned by Singh (2022):

- **Balanced Technological Advancement.** While maintaining a focus on sustainability, Anarres could benefit from embracing technological advancements that improve living standards and innovation. Integrating appropriate technologies that align with ecological principles could address the harsh living conditions and resource scarcity without compromising environmental values.
- **Enhanced Efficiency and Incentives.** Addressing the inefficiencies and lack of personal incentives within the communal ownership system of Anarres could improve productivity and resource management. Implementing a balanced approach that encourages individual initiative while maintaining communal values might foster a more dynamic and motivated society.
- **Holistic Approach to Well-Being.** Emphasizing a holistic approach to well-being that integrates mental, physical, and spiritual health could enhance the overall quality of life on Anarres. Providing access to more diverse forms of cultural activities, education, and healthcare could contribute to a balanced and fulfilling existence.
- **Improved Coordination and Governance.** Strengthening the coordination and coherence of decentralized governance in Anarres could mitigate the challenges of fragmented policies and local conflicts. Establishing mechanisms for better communication and collaboration among different communities and decision-making bodies could enhance the overall effectiveness of governance.
- **Social and Cultural Adaptability.** Encouraging a culture of adaptability and resilience could help Anarres better navigate the complexities and contradictions within its society. Promoting open dialogue, continuous learning, and flexibility in social and cultural practices could foster a more inclusive and dynamic community.

10.6 On Brave New Worlds (Redux)

Le Guin (1974) explores ecotopian ideals through the portrayal of Anarres, a society that emphasizes ecological harmony, social equality, and communal values. This précis examined concepts such as communal ownership, decentralized governance, and sustainable living, underscoring their relevance to today's pressing societal issues. However, it also acknowledged Le Guin's cautionary stance, reflecting the concerns that Mumford (1965) raises regarding the technical and practical challenges of realizing such ideals. Ultimately, the précis encourages readers to consider Singh's (2022) concept of "utopias of the third kind," which advocates for adaptable, resilient, and realistic visions of the future. By analyzing the complexities and contradictions within both Anarres and Urras, we gain a fuller appreciation of the need to balance ecological sustainability, social equity, and technological progress.

References

Le Guin U (1974) The dispossessed. Harper & Row
Mumford L (1965) Utopia, the city, and the machine. Daedalus 94(2):271–292
Serrat O (2025) Building sustainable futures: key themes in ecotopian visions. Unpublished manuscript, Georgetown University. https://www.researchgate.net/publication/388651507_Building_Sustainable_Futures_Key_Themes_in_Ecotopian_Visions
Singh V (2022) Utopias of the third kind. PM Press

Chapter 11
From Deforestation to Gaia: A Journey of Reflection and Respect

Abstract This précis delves into the environmental devastation caused by logging and monoculture in Côte d'Ivoire and the inspiring journey of discovering René Dubos's and James Lovelock's environmental philosophies. The narrative transitions from personal reflection to a broader call emphasizing the importance of transcending human-centered thinking and developing a respectful relationship with nature.

11.1 Wheels of Destruction: The *Grumiers*

Growing up in Côte d'Ivoire, I could not fail to notice the deforestation reshaping the landscape. Massive *grumiers*—16- to 22-wheel lumber trucks—rumbled through narrow laterite roads, kicking up clouds of reddish-brown dust and scattering iron and aluminum oxide pellets onto the verges. These behemoths hauled trees of such girth that they could only carry three, and sometimes just two, at a time. When one appeared ahead, my father would pull over and let the dust settle before we resumed our journey.

The *grumiers* unloaded their cargo—chainsawed trunks of African Mahogany (*Khaya spp.*) into the Bandama River. African Ebony (*Diospyros crassiflora*), Ceiba (*Ceiba pentandra*), Iroko (*Milicia excelsa*), and Sapele (*Entandrophragma cylindricum*) were also discharged. A watery conveyor belt, the Bandama delivered the dead harvest to the coast, destined for shipment overseas to become luxury furniture, flooring and paneling, veneers and plywood, or even parts of musical instruments.

At the time of its independence in 1960, nearly 50% of Côte d'Ivoire's land area was covered by dense tropical rainforest, predominantly located in the southern half of the country, where the climate is more humid. However, over the following decades, particularly during the 1960s and 1970s, over 80% of the nation's forest cover vanished, a consequence of rampant timber harvesting. The destruction deprived countless animal and plant species of their natural habitats and severely impacted indigenous communities. The hunter-gatherer Bakwé Pygmies were among the most affected, with their population reduced to only 20,000 in 2025 (International Mission Board 2025; Serrat 2021a). Stripped of its once-emerald canopy, Côte

O. Serrat, *Anthropogenic Solutions for Climate Change*,
https://doi.org/10.1007/978-981-96-8654-4_11

d'Ivoire is now the world's largest producer of cocoa beans, trapped in a monoculture economy subject to market volatility. In the eponymous Ivory Coast, the forest elephant population has plummeted to fewer than 225, a tragic decline from the estimated 3000–5000 that roamed the region a century ago (Kouakou et al. 2020).

11.2 From Deforestation to Respect: A Personal Awakening

As widespread forest loss ravaged neighboring countries such as Ghana and Nigeria, and indeed entire regions worldwide (e.g., the Amazon rainforest, Central Africa, Southeast Asia), the disquiet I felt as a child over the fate of Côte d'Ivoire's rainforest drew me to the works of René Dubos (1901–1982). A French-American microbiologist and environmentalist, Dubos gained renown for advocating a balance between humanity and nature. A quote I encountered left a lasting impression: "Our salvation depends upon our ability to create a religion of nature" (Dubos and Ward 1983, p. 24).

Dubos's Pulitzer Prize-winning book, *So Human an Animal* (1968), explored transformative themes including the humanization of science, environmental stewardship, the interconnectedness of life, and innovation driven by optimism. His philosophy, encapsulated in the adage he helped popularize (and some argue he actually coined)—"think globally, act locally"—underscored humanity's delicate yet vital connection to nature. *So Human an Animal* (1968) holds a place of honor on my bookshelf, serving as a constant reminder that another world is not merely imaginable: it is entirely possible.

Do not dismiss this notion—it is more attainable than it may seem. And even if the vision of another world should remain out of reach, countless opportunities still abound to lead more meaningful and fulfilling lives. In this context, while Dubos set the foundation for responsible care of our natural surroundings, James Lovelock (1919–2022), an English environmentalist and futurist, expanded this ecotopian vision to encompass the entire planet. Lovelock (1979) conceptualized Earth's systems as a dynamic, interconnected organism, which, at the suggestion of William Golding—the acclaimed author of *Lord of the Flies* (1954)—he named Gaia after the Greek goddess of the Earth. Looking back on my childhood, Dubos's compelling plea for people to cultivate respect for the environment resonated with me. Over time, my journey took me across more than a dozen countries in Asia and the Pacific, where I dedicated myself to advancing prosperity, inclusivity, resilience, and—most importantly—actionable and transformative sustainability (Serrat 2021b).

11.3 Reimagining the Future: Transcending Human-Centered Thinking

Crises may ignite change, but true transformation in our precarious times requires more than mere adaptation—it demands an unwavering commitment to envisioning and creating a sustainable future. Realizing this vision calls for global collaboration, widespread education and awareness, bold innovation, economic reform, cultural evolution, principled leadership, and active community participation. Yet, drawing inspiration from Dubos, I believe that transcending human-centered thinking is crucial; it must serve as the foundation upon which these future-minded efforts are built. This raises a pivotal question: how do we embark on this vital journey?

In answering this, *An Ecotopian Lexicon* (Schneider-Mayerson and Bellamy 2019) offers ideas by introducing 30 loanwords from various cultures and settings to address environmental issues. Two terms—*fotminne* and *watershed discipleship*—remind us of the spiritual responsibility to live symbiotically with our natural environment. *Fotminne*, a Swedish term meaning "foot memory," refers to the embodied memory of walking through a landscape. It emphasizes the profound connection between humans and the natural world and encourages us to honor the places we have traversed. This concept reminds us that every step we take leaves an imprint—not only on the land itself but also in our own sense of identity and belonging. It challenges us to tread lightly, to appreciate the stories held within the earth beneath our feet, and to pass on these memories as part of a collective cultural and ecological heritage. *Watershed discipleship* combines bioregionalism, ecological theology, and faith-based. It focuses on the bioregional context, urging people to reconnect with their local environments; explicitly acknowledges the current watershed moment of our ecological crisis; and calls for discipleship, encouraging individuals and communities to become stewards of their surroundings. Watersheds connect us: they provide foundational, practical, and scalable units of care. In unison, *fotminne* and *watershed discipleship* stress the urgent need to move beyond human-centered perspectives to develop a respectful relationship with nature.

11.4 Beyond the Deluge: Regaining Fervor

Alongside deforestation, a range of environmental challenges—such as climate change, air, soil, and water pollution, biodiversity loss, and unsustainable land use—are increasingly threatening ecosystems and biodiversity. Meanwhile, ocean acidification, plastic pollution, rising sea levels, and the depletion of natural resources are further intensifying the strain on Earth's delicate balance. In response, contemporary ecotopian, dystopian, and utopian literature has witnessed a revival of religious and quasi-religious themes, reflecting humanity's moral and spiritual grappling with the ecological crises.

Echoing the memorable observation by Dubos and Ward (1983) about humanity's need to establish a religion centered on nature, Margaret Atwood's *The Year of the Flood* (2009) vividly portrays an eco-religious group devoted to reverence for nature and harmonious coexistence with the Earth. Known as "God's Gardeners," the members of the group embody their commitment to environmental stewardship through organic farming, vegetarianism, and sustainable practices. Central to their teachings is a belief in the "Waterless Flood," an impending cataclysm they foresee will upend humanity's existence. Their lifestyle blends ecological awareness with spirituality, forging a unique fusion of faith and environmentalism. A key cultural element is *The God's Gardeners Oral Hymnbook*, a fictional collection of hymns written in poetic form. These hymns not only encapsulate their philosophy of sustainable living but also guide their values, shape their actions, and foster a collective identity, serving as both a moral compass and a unifying force.

Equipped with our newly expanded ecotopian lexicon, we would benefit greatly from drawing inspiration from the visionary works of thinkers such as Atwood (2009), Dubos (1968), and Lovelock (1979). Each contributes catalytic perspectives on the essential need for humanity to respect the Earth. Dubos emphasizes the moral responsibility to act locally while thinking globally, advocating for an ethical engagement with nature to ensure humanity's well-being. Lovelock's Gaia theory expands this vision to a planetary scale, reminding us that Earth itself is a living, interconnected organism that demands our care and protection. Atwood, through her portrayal of "God's Gardeners," illustrates the transformative potential of blending spiritual reverence with ecological stewardship, urging us to reimagine our relationship with the planet "[T]o sense the wholeness of the Whole, Not only with a thinking Brain, But with a swift and ardent Soul" (Atwood 2009, p. 236).

Essentially, as seen from Lockyer and Veteto (2013), Atwood (2009), Dubos (1968), and Lovelock (1979) signify that sustainability cannot be achieved by fixating solely on the intertwined environmental and social crises we face. Instead, as Lockyer and Veteto also suggest, the time has come to prioritize solution-driven ideas—many of which have already emerged from the grassroots—and to move beyond strategies that focus predominantly on highlighting problems. Together, their perspectives form a powerful call to action: to safeguard and honor the delicate balance of life on Earth, not only to ensure the survival of our species but also out of respect for the intricate and fragile web of existence that sustains all. Humility is the springboard of transformation. Dreams ignite possibilities. Resolve turns vision into action. Let us uphold our shared responsibility to care for our planet, our "Mother Earth": *Pachamama*!

References

Atwood M (2009) The year of the flood. Vintage

Dubos R (1968) So human an animal: how we are shaped by surroundings and events. Charles Scribner's Sons

Dubos R, Ward B (1983) Only one Earth: the care and maintenance of a small planet. W. W. Norton & Company. (Original work published 1972)

Golding W (1954) Lord of the flies. Faber and Faber

International Mission Board (2025) Bakwé of Côte d'Ivoire. PeopleGroups.org. https://www.peoplegroups.org/explore/groupdetails.aspx?peid=13041

Kouakou J-L, Gonedelé Bi S, Bitty EA, Kouakou C, Yao AK, Kassé KB, Ouattara S (2020) Ivory Coast without ivory: massive extinction of African forest elephants in Côte d'Ivoire. PLOS ONE 15(10):e0232993

Lockyer J, Veteto J (eds) (2013) Environmental anthropology engaging ecotopia: bioregionalism, permaculture, and ecovillages. Berghahn Books

Lovelock J (1979) Gaia: a new look at life on Earth. Oxford University Press

Schneider-Mayerson M, Bellamy B (eds) (2019) An ecotopian lexicon. University of Minnesota Press

Serrat O (2021a) Minority population analysis: the Aeta of the Philippines. In: Leading solutions: essays in business psychology. Springer, pp 267–271

Serrat O (2021b) Life-markers and personal values. In: Leading solutions: essays in business psychology. Springer, pp 55–56

Chapter 12
Mottainai: Integrating Waste Mindfulness into Our Ecological Vocabulary

Abstract This précis advocates for the inclusion of *mottainai*, a Japanese concept embodying waste mindfulness and respect for resources, into our ecological vocabulary. By adopting *mottainai*, we can deepen our cultural understanding of sustainability, foster mindful consumption, and promote holistic ecological practices. The proposal highlights its relevance in addressing contemporary environmental challenges and enriching global ecological discourse.

Pronunciation: moht-tai-nai (mɒt.taɪ.naɪ)
Part of Speech: Adjective (Interjection)
Provenance: Japanese, mainly Buddhist
Example: Honoring the ethos of *mottainai*, the town breathed new life into an abandoned warehouse, reimagining it as a dynamic hub for community engagement. With reclaimed wood and repurposed materials, local residents crafted a space dedicated to skill-sharing workshops, second-hand markets, and environmental education programs—cultivating a shared appreciation for sustainability and intentional living.

What if we cherished every resource, no matter how small, and utilized it to its fullest potential? *Mottainai* (Japanese: もったいない or 勿体無い) is a traditional Japanese expression rooted in Buddhist philosophy and Shintoism, Japan's oldest spiritual tradition: it conveys a sense of regret over wastefulness and a recognition of the intrinsic value in resources, objects, and experiences. The word encompasses ecotopian ideals of conservation, frugality, resourcefulness, and sustainability, and summons mindfulness, respect, and gratitude for the materials and energy that sustain life. Its etymology and historical context reflect a blend of cultural, spiritual, and practical values, evolving over centuries to address both environmental and societal challenges.

To truly understand *mottainai*, it is necessary to delve into its linguistic origins. The word *mottainai* consists of two parts: *mottai* (Japanese: もったい or 勿体) and the suffix *nai* (Japanese: ない or 無い). In classical Japanese, *mottai* referred to the essence, importance, sanctity, or substance of something, while *nai* translates to "without" or "lacking." In artful combination, therefore, the words convey the sentiment of something being wasted or not fully utilized, evoking a deep sense of regret.

O. Serrat, *Anthropogenic Solutions for Climate Change*,
https://doi.org/10.1007/978-981-96-8654-4_12

The term can be interpreted as "What a waste!" or, more fittingly, the old saying, "Waste not, want not." This concept is firmly anchored in Buddhist teachings, which call attention to the impermanence and interconnectedness of all things (Serrat 2024a, b). Wastefulness is seen as a failure to recognize and appreciate the inherent value and effort embedded in resources, encouraging an attitude of mindfulness, gratitude, and respect.

Throughout history, *mottainai* has been a cornerstone of Japanese culture. During the Edo period (1603–1868), Japan's isolationist policy restricted foreign trade to safeguard the nation's cultural and political systems while minimizing external influences. This policy necessitated self-sufficiency and efficient resource use, solidifying *mottainai* as a core philosophy for resource management, environmental consciousness, and sustainable living. People were compelled to make the most of limited resources, embedding *mottainai* into daily practices. In modern times, the term has gained renewed prominence as a rallying cry for environmental conservation, advocating for actions like recycling, reducing waste, and valuing everyday objects.

Beyond its practical applications, *mottainai* holds weighty cultural significance in Japanese life. It represents not only practical resource management but also boasts a spiritual dimension. It motivates individuals to live with humility, fostering gratitude and responsibility. This perspective permeates daily activities, from preparing meals thoughtfully to repairing and reusing items, serving as a reminder to respect and protect the natural world.

In our increasingly globalized world, the principles of *mottainai* have become quite relevant as communities and nations grapple with the challenges of climate change, pollution, and resource depletion. The consequences of wastefulness and environmental degradation are felt on a global scale, underscoring the importance of sustainable practices. Embracing *mottainai* offers a meaningful path toward building a resilient future, with practical applications including community involvement in environmental initiatives, adopting eco-friendly materials, conserving energy, and embracing minimalism. By internalizing the spirit of *mottainai*, individuals and societies can cultivate a sustainable and harmonious existence, while honoring the quintessential interrelation of all life.

Practical ways to apply *mottainai* in daily life include participating in environmental initiatives like clean-up events, tree planting, or sustainability workshops, and spreading the message to inspire collective action. Eco-friendly habits, such as using reusable bags, containers, and bottles, choosing minimally packaged products, and supporting sustainable brands, help reduce waste. Energy conservation practices like turning off unused appliances, using energy-efficient devices, and switching to renewable energy sources are *de rigueur*. Growing vegetables and herbs, even in small containers, reduces the carbon footprint of food transportation while fostering a connection to the earth. Mindful consumption encourages buying only what is needed and selecting high-quality, durable items. A minimalist lifestyle, achieved by decluttering, recycling, and donating unneeded items, along with meal planning and composting, helps reduce waste. Repairing and reusing items, repurposing materials, opting for sustainable transportation (i.e., walking, biking, public transit), and conserving water through fixes and efficient fixtures are practical *mottainai* principles anyone can adopt.

Mottainai extends far beyond individual practices; as a collective philosophy it has the potential to shape modern society through transformative avenues. Art and media are powerful tools where artists, filmmakers, and writers can use their platforms to highlight the concept and back sustainable practices. Corporate responsibility can lead businesses to adopt sustainable practices and promote a culture of resourcefulness with plentiful applications. Cultural celebrations through festivals and events can emphasize sustainability and gratitude. Education and awareness are crucial, and schools, universities, and community organizations can incorporate *mottainai* into their curricula to raise awareness. Policy and governance can also play a crucial role, with governments promoting *mottainai* through policies and regulations.

Outside Japan, the concept of *mottainai* has been championed by Wangarĩ Maathai, the Kenyan environmentalist. In 1977, she established the Green Belt Movement (https://www.greenbeltmovement.org/), a groundbreaking initiative dedicated to environmental conservation through tree planting, combating deforestation, and restoring ecosystems. The movement aims to uplift communities—particularly women—while promoting sustainable livelihoods. Additionally, it advocates for women's rights, democratic accountability, and grassroots community empowerment. Maathai was awarded the Nobel Peace Prize in 2004. In 2005, during a visit to Japan, she came across the term *mottainai* and integrated it into her global environmental advocacy. By advancing the principles of "reduce, reuse, recycle" alongside a fourth "R"—respect—she amplified awareness of *mottainai* practices to help tackle pressing global environmental challenges.

In closing: Japanese: 地球上のあらゆる場所で、もったいないの精神を取り入れることで、私たちはより持続可能で調和のとれた関係を育むことができるのです. (By embracing the spirit of *mottainai* around the world, we can foster a more sustainable and harmonious relationship with our planet.) The enduring philosophy of *mottainai* provides a valuable framework for mindful living, resource conservation, and environmental stewardship. It mitigates excessive consumption and guides us toward a greener future. I propose that the concept of *mottainai* be considered for inclusion in a future edition of Schneider-Mayerson and Bellamy (2019). The lexicon already includes the concept of *shikata ga nai* (Japanese: 仕方がない or しかたがない), an expression that translates to "it cannot be helped" or "it is what it is." *Shikata ga nai* reflects a sense of acceptance and adaptability in the face of unavoidable circumstances. When applied to the ecological crisis, that concept conveys the importance of recognizing the realities we face and directing our efforts toward actionable solutions, rather than becoming immobilized by aspects beyond our control. The rationale for including *mottainai* alongside *shikata ga nai* lies in their complementary perspectives, blending realism with responsibility.

References

Schneider-Mayerson M, Bellamy B (eds) (2019) An ecotopian lexicon. University of Minnesota Press

Serrat O (2024a) Knowing that and knowing how: must it be either/or? Unpublished manuscript, Georgetown University. https://www.researchgate.net/publication/379514661_Knowing_That_and_Knowing_How_Must_It_Be_EitherOr

Serrat O (2024b) Bridging worlds: comparative epistemology and the mind–body debate. Unpublished manuscript, Georgetown University. https://www.researchgate.net/publication/380611280_Bridging_Worlds_Comparative_Epistemology_and_the_Mind-Body_Problem

Chapter 13
The Man Who Planted Trees: Ecotopian Ideals in Giono's (1953) Storytelling

Abstract *The Man Who Planted Trees* (1953) by Jean Giono is a profound allegorical tale. This précis examines the short story as a foundational work of ecotopian literature, illustrating how perseverance and ecological stewardship can restore barren landscapes into thriving ecosystems. Through an analysis of themes such as renewal, sustainability, and human–nature symbiosis, the précis positions the narrative as both a literary masterpiece and an enduring call to environmental action.

13.1 Sowing Seeds of Hope

Long before Ernest Callenbach's *Ecotopia* (1975) captivated audiences with its vision of sustainable utopias, Jean Giono (1895–1970) quietly pioneered the genre with his allegorical 1953 short story *The Man Who Planted Trees.*[1] A celebrated French author known for his depictions of Haute-Provence's landscapes, Giono's outlook was shaped by his harrowing experiences during the Great War. Drafted at the onset of the conflict, he was thrust into the Battle of Verdun (February–December 1916)—grimly known as *La Fournaise* (The Furnace). That apocalyptic confrontation wrought unspeakable devastation, claiming more than 710,000 French and German lives in less than a year. The carnage Giono witnessed—first in Verdun, then through the war's remainder—forged his lifelong commitment to pacifism, permeating his work with themes of hope, ecological balance, and the enduring power of renewal.

[1] Samuel Butler's *Erewhon* (1872) and William Morris's *News from Nowhere* (1890) are often labeled as ecotopian literature but this is misleading. *Erewhon* critiques Victorian society, focusing on morality, religion, and industrialization, with environmental themes serving as a backdrop. *News from Nowhere* envisions a socialist utopia, likewise critiquing capitalism and industrialism, but its pastoral setting is secondary to its social critique.

O. Serrat, *Anthropogenic Solutions for Climate Change*,
https://doi.org/10.1007/978-981-96-8654-4_13

13.2 A Sheperd's Miracle

The creation of *The Man Who Planted Trees* is as remarkable as the message it conveys. Written in French under the title *L'homme qui plantait des arbres*, Giono's crafted the piece—spanning approximately 4000 words—in just two days. He submitted it in 1953 as an entry for a *Reader's Digest* competition seeking essays about "The Most Unforgettable Character I've Met." Giono, however, performed an imaginative leap. Instead of recounting the story of a real-life individual, he chose to craft a fictional narrative, weaving the tale of Elzéard Bouffier, a humble yet determined shepherd. With unwavering dedication, Bouffier single-handedly transforms a lifeless valley into a thriving, lush environment. Giono's definition of an unforgettable character is timeless. In his view:

> For a human character to reveal truly exceptional qualities, one must have the good fortune to be able to observe its performance over many years. If this performance is devoid of all egoism, if its guiding motive is unparallelled generosity, if it is absolutely certain that there is no thought of recompense and that, in addition, it has left its visible mark upon the earth, then there can be no mistake. (Giono 1985, p. 3)

The story unfolds through the perspective of an unnamed traveler—a surrogate for Giono himself, perhaps—who stumbles upon Bouffier's purposeful mission. Initially encountering him in a starkly barren landscape, the traveler watches as Bouffier methodically sows acorns with quiet determination. Over time, each subsequent visit reveals remarkable changes. Spanning decades from 1913 until Bouffier's death in 1947, the once lifeless land blossoms again: dried-up streams resume their flow, wildlife returns to reclaims its habitat, and abandoned villages spring back to life as settlers rebuild vibrant communities. Through the narrator's recollections, we observe the ripple effects of one man's altruistic commitment, showcasing how perseverance can breathe life into desolation:

> When I reflect that one man, armed only with his own physical and moral resources, was able to cause this land of Canaan to spring from the wasteland, I am convinced that in spite of everything, humanity is admirable. (Giono 1985, pp. 34–35)

13.3 Giono's Gift to Humankind

Reader's Digest declined *The Man Who Planted Trees*, deeming it ineligible since the competition required essays about real-life individuals. Undeterred, Giono released his tale into the public domain, calling it as a gift to humankind. The tale first appeared in *Vogue* under the title *The Man Who Planted Hope and Grew Happiness*. Over time, English translations brought the narrative to a global audience, garnering international acclaim. Frédéric Back (1987) adapted it into an animated short film, with a runtime of approximately 30 minutes. Today, *The Man Who Planted Trees* is regarded as a timeless environmental manifesto. The 2005 twentieth-anniversary edition of Chelsea Green Publishing's 1985 release highlights the tale's enduring impact. This commemorative edition features a foreword by Nobel Laureate Wangari Maathai, whose Green Belt Movement has inspired hope and renewal in communities across Kenya through the planting of millions of trees. Maathai eloquently underscores the story's universal message, writing:

> The Green Belt Movement and *The Man Who Planted Trees* make a very important point: you do not need a diploma to make a difference; everyone is qualified to save the planet. (Giono 1985, p. ix)

13.4 Ecotopian Themes in *The Man Who Planted Trees*

The Man Who Planted Trees embodies a wealth of universal themes, solidifying its status as a foundational work in discussions on ecotopian visions—ideals of societies flourishing in harmony with nature through sustainable practices and communal well-being (Serrat 2024). The story presents ecological restoration as a pathway to healing both the environment and society (viz., human–nature symbiosis). Bouffier's steadfast efforts illustrate the principles of sustainability, demonstrating how small, consistent actions can ignite long-term environmental regeneration (viz., restoration and sustainability). As the once-barren landscape prospers anew, abandoned villages and communities are revitalized, underscoring the vital connection between ecological health and human prosperity (viz., community and prosperity). The narrative also conveys hopeful resilience, exemplifying how determination can overcome adversity to rejuvenate ecosystems and societies (viz., hope and resilience). Through Bouffier's solitary yet transformative work, the story inspires collective action, highlighting how individual efforts can catalyze broader movements for climate action and sustainable living (viz., catalyst for environmental action). Ultimately, Giono offers an ecotopian vision of a harmonious future—one that provides both optimism and a roadmap for confronting today's environmental challenges with creativity and resolve (viz., ecotopian vision).

The ecotopian tone of *The Man Who Planted Trees*, marked by themes of harmony, sustainability, aspiration, and legacy, reflects the ideals of ecotopian thought. Giono suggests that even the most modest acts can precipitate meaningful change. In an era plagued by climate challenges, deforestation, and ecological instability, his tale reminds us that humankind has the ability to commit, create, and connect with the Earth. Let it inspire each of us to act—plant a tree, cultivate gardens, or promote sustainable living. Every step, however small, brings us closer to a future where nature and human prosperity thrive together.

References

Back F (Director) (1987) The man who planted trees [Short film]. Société Radio-Canada.

Butler S (1872) Erewhon: or over the range. Trübner & Co.

Callenbach E (1975) Ecotopia: the notebooks and reports of William Weston. Bantam Books

Giono J (1985) The man who planted trees (D. Wilson, Trans.). Chelsea Green Publishing. (Original work published 1953)

Morris W (1890) News from Nowhere: or an epoch of rest. Reeves & Turner

Serrat O (2024) Building sustainable futures: key themes in ecotopian visions. Unpublished manuscript, Georgetown University. https://www.researchgate.net/publication/388651507_Building_Sustainable_Futures_Key_Themes_in_Ecotopian_Visions

Part V
Environmental Migration

Chapter 14
Environmental Migration in the Anthropocene: Perspectives on the Relocation of Indonesia's Capital City to Eastern Borneo

Abstract Climate change is a driver of human mobility: it is expected to increase the displacement of populations. This précis casts environmental and socio-economic perspectives on the relocation of Indonesia's capital city from Java to eastern Borneo, the first instance of large-scale, anticipatory, and managed environmental migration in the Anthropocene.

14.1 The Setting

14.1.1 A Brief History of Jakarta

Jakarta has been inhabited since the fourth century CE: at the time, a small Indianized settlement and trading hub of Tarumanagara—the first recorded kingdom in western Java—thrived there because of the area's strategic location at the mouth of the Ciliwung River (Brittanica 2024a). The original settlement, Sunda Kepala, connected the Indonesian archipelago with the territory stretching between India and China; it was renamed Jakarta (Sundanese: Jayakerta or "Glorious Fortress") in 1527 after the Sultanate of Demak, a Javanese Muslim state located on Java's north coast, repulsed a Portuguese attack on the harbor (Brittanica 2024a). It was in the early seventeenth century that Jakarta became a crucial trading outpost. Captured and razed by the Dutch in 1619, then renamed Batavia, the outpost then served as the *de facto* capital of the Dutch (or Netherlands) East Indies, a copycat Amsterdam with canals, gabled merchants' houses, and warehouses (Brittanica 2024b). Batavia and its cosmopolitan population boomed in a mosquito-infested, swampy expanse of opulent mansions and crowded slums. In 1942, the Dutch East Indies were overrun by Japan, looking to secure raw materials (e.g., oil, rubber, tin) to fuel its war economy: Batavia was occupied and renamed Jakarta. Indonesia proclaimed its independence when the collapse of Japan was confirmed in 1945. In 1949, after attempts to reclaim the Dutch East Indies had failed, the Netherlands reluctantly acceded to their independence and Jakarta became the capital of the new Republic of Indonesia (Brittanica 2024b). Indonesia achieved independence amid general

O. Serrat, *Anthropogenic Solutions for Climate Change*,
https://doi.org/10.1007/978-981-96-8654-4_14

political turmoil, economic instability, and social inequality. The difficulties that the new Government of Indonesia faced were born of the pernicious effects of racist colonialism (e.g., coerced cultivation of cash crops; slave, coolie, convict, and corvée labor); famine and forced labor on economic development and defense projects during the Japanese occupation from 1942; and the newly-felt need to unite ethnic and religious groups under a common national identity, which led to internal upheavals and communal violence (Wikipedia 2024a). With the advent of large-scale construction and monumental projects under President Sukarno in the 1960s, Jakarta experienced explosive growth: infrastructure was modernized, office towers rose, and the city became one of the largest metropolises of Southeast Asia. In 1950, Jakarta's population had numbered about 1.5 million people; by 1970, it had reached 3.9 million. Jakarta's population stood at 11.4 million in 2023; it is projected to reach 13.7 million in 2035 (World Population Review 2024). With close to 279 million people in 2024 living on 17,508 islands, the largest of which in descending order are New Guinea (shared with Papua New Guinea), Borneo (shared with Brunei Darussalam and Malaysia), Sumatra, Sulawesi, and Java, Indonesia is the fourth most populous country in the world after China, India, and the United States (World Population Review 2024).

14.1.2 Environmental Challenges Facing Jakarta

All the time more, environmental challenges associated with overconcentration, overdevelopment, overpopulation, and location in a deltaic floodplain have come to threaten Jakarta's livability and sustainability (Syaban and Appiah-Opoku 2023). A recent study ranking the world's 576 largest urban centers on their exposure to a range of environmental and climate-related threats identified Jakarta as the worst-performing city (Nichols 2021).

14.1.2.1 Traffic Congestion

The metropolitan area of Jakarta stretches over 6392 square kilometers (2468 square miles) (Brittanica 2024a). However, the city proper was built on a deltaic floodplain through which 13 low-lying rivers drain and only covers 664 square kilometers (256 square miles) (Brittanica 2024a). Because of rapid urbanization and economic growth, Jakarta had a population density of about 16,000 people per square kilometer (about 42,000 people per square mile) in 2022 (Brittanica 2024a). Office buildings, residential areas, and shopping centers are generally built along highways and do not connect to public transportation (Van Mead 2016). A high population density has brought with it many private cars and motorcycles that, over a limited road network, exacerbate traffic congestion despite the construction of flyovers, toll roads, bus lanes, and mass rapid transit systems (Van Mead 2016). Jakarta is one of the most congested cities in the world: spending four hours in traffic is not unusual (Van Mead 2016).

14.1.2.2 Air Pollution

Air pollution is one of the most pressing environmental issues in Jakarta: the main sources are vehicle exhaust emissions, coal combustion from power plants, open waste burning, household emissions, construction and road dust, industrial emissions from manufacturing, and resuspended soil particles (Earth5R 2020). These sources emit particulate matter (PM), carbon monoxide (CO), ozone (O_3), nitrogen dioxide (NO_2) and sulfur dioxide (SO_2), pollutants that can cause respiratory diseases, cardiovascular problems, cancer, and premature death (World Health Organization 2024). A study by Greenpeace Indonesia, reported in Earth5R (2020), estimated that at least 7390 inhabitants of Jakarta die every year due to high levels of PM2.5 (particulate matter with a diameter less than 2.5 microns) and almost 2000 babies are born with low birth weights for the same reason. Air pollution also contributes to global warming.

14.1.2.3 River Pollution

River pollution is another major environmental challenge in Jakarta: it affects the water quality and ecosystem of the city and its surroundings. Thirteen rivers flow through Jakarta before emptying in Jakarta Bay: listing alphabetically, they are the Angke, Baru Barat, Baru Timur, Buaran, Cakung, Ciliwung, Cipinang, Grogol, Jatikramat, Krukut, Mookervaart, Pesanggrahan, and Sunter Rivers. A study by the Ministry of National Development Planning (BAPPENAS), referred to by the Jakarta Post (2018), reported that 96% of river water in the city is severely polluted. Of the 13 rivers, according to Earth5R (2020), only Krukut River's water is clean enough to be treated for potable water; the 12 other rivers are contaminated by plastic and littering but also household, liquid, industrial, and agricultural waste (Gokkon 2022). The most polluted river in Jakarta is the Citarum River: black, smelly, and covered with trash, it has been dubbed the most polluted river in the world (Hutton 2013). Many people leaving along the Citarum River suffer from contact rashes, dermatitis, and intestinal problems but delays in child development, chronic bronchitis, renal failure, and high incidence of tumors have also been reported (Carrubba 2020).

14.1.2.4 Flooding

Flooding affects the lives and livelihoods of millions of people in Jakarta: the city's elevation averages as mere 8 meters (about 26 feet) above sea level. Jakarta is regularly hit by floods during the monsoon season, which lasts from October to March.

The floods are caused by highly intense rainfall; poor drainage through river channels canals, and waterways that have narrowed or become periodically clogged with sediment and trash; land subsidence; changes in land cover that have replaced open spaces; and deforestation in Jakarta's hinterlands. The worst flood in Jakarta's history occurred in 2007 when more than 70% of the city was inundated, affecting more than 2.5 million people and causing over $570 million in damages (NASA Earth Observatory 2019; Wikipedia 2024b). Adding to damage to infrastructure, the floods recurrently disrupt transportation, communication, education, and business activities; they also pose health risks to the population from waterborne diseases, skin infections, and respiratory problems.

14.1.2.5 Land Subsidence

Land subsidence affects the stability and safety of Jakarta with economic, environmental, infrastructural, and social impacts. Indonesia's capital city is built on swampy land characterized by a high water table and soft soil. Jakarta's rapid population growth, combined with inadequate infrastructure and poor urban planning, has led to increased groundwater extraction (Renaldi 2022). The residents of Jakarta use their own wells to obtain freshwater from underground: as groundwater is pumped out, the alluvial plain's soft soil loses its support and compacts (Aqil 2023). Land subsidence has also been aggravated by the lack of green space, extensive conversion of agricultural areas, the load of constructions, and because asphalt and concrete prevent the absorption of water into the soil (Aqil 2023). Drained of its vital groundwater reservoirs, Jakarta has sunk by 4 meters over the last 30 years; some parts of the city have sunk by 2.5 meters in the last 10 years (Lin and Hidayat 2018). Land subsidence has caused damage to bridges, buildings, roads, and other infrastructure, and increased the risk of flooding and coastal erosion. Subsidence can also alter the stress distribution in the earth's surface, potentially inducing seismic activity because Jakarta lies on the Ring of Fire, the belt of active volcanoes, volcanic arcs, and tectonic plate boundaries that frame the Pacific Ocean (Brittanica 2024c; Gambolati and Teatini 2021).

14.1.2.6 Climate Change

Climate change is the long-term change in the average weather patterns of the Earth, caused by the increase in greenhouse gas emissions from human activities such as burning fossil fuels, deforestation, and agriculture (National Aeronautics and Space Administration 2024). Climate change has global and local impacts including higher temperatures, more frequent precipitation events and storms, faster rates of ocean warming, and sea-level rise (National Aeronautics and Space Administration 2024). Climate change is a "force multiplier" that looms over Jakarta: it may exacerbate the environmental issues currently affecting Jakarta and give a free rein to new ones, such as heat waves, droughts, and storms. About 40% of Jakarta is already below sea level: because global sea levels are rising by 3.3 millimeters per year, Jakarta is highly exposed and likely to be evermore affected by coastal erosion and groundwater and soil salinization; inland penetration of tropical storm surges; inundation; sea-level rise; and shifting sediment regimes (Renaldi 2022). These hydrometeorological hazards could damage agriculture and freshwater supplies in the vicinity of Jakarta, unsettle human systems, downgrade the value of property, and destroy infrastructure to accelerate migration (Hauer et al. 2019). It is estimated that one-third of the city could be submerged by 2050 (Associated Press 2022).

14.1.2.7 The Government of Indonesia's Response

The environmental challenges that threaten Jakarta are interrelated: they are the upshot of the aggravated disconnect between natural (e.g., geography, geology, hydrology, meteorology) and human factors (e.g., culture, demography, economy, politics) and pose serious risks to the city's sustainability. Climate change may compound the environmental challenges facing Jakarta and a quarter of the city could be underwater by 2050 (Lin and Hidayat 2018). (Jakarta is also close to two active volcanoes, Gunung Gede and Krakatau. Gunung Gede is but 70 kilometers or 43 miles away.) "We cannot continue to burden Jakarta and Java any more in terms of population density, congestion, pollution, and water resources," President Joko Widodo declared in 2019 in recognition that the risk of remaining might become too high (Aditya et al. 2019). (President Widodo was governor of Jakarta from 2012 to 2014 and mayor of Surakarta from 2005 to 2012.) Faced with a perfect storm of climate and environmental challenges, the Government of Indonesia then decided to relocate to a new capital city about 1500 kilometers (930 miles) away in the province of East Kalimantan, in the eastern part of the island of Borneo, between Balikpapan and the provincial capital of Samarinda (Serrat 2024). Nusantara, a purpose-built, smart, and "sustainable forest city," will cover about 256,000 hectares (990 square miles) and cost an estimated $32 billion with 80% participation from the private sector. (Nusantara is an old Javanese term that means "archipelago.") The bill of the Law on State Capital was promptly passed by Indonesia's parliament on January 18, 2022 and the law became effective on February 15, 2022 (Cabinet Secretariat of the Republic of Indonesia 2022). On April 18, 2022, the Government of Indonesia formally established the Nusantara Capital City Authority (Indonesian: Otorita Ibu Kota Nusantara, abbreviated as IKN) as a cabinet level-agency working directly under the President of Indonesia to manage and govern Nusantara (Nusantara 2024). The permanent constitution of the Authority, Presidential Regulation No. 62/2022, was published on May 4, 2022, although the date of the decree was backdated to April 18, 2022 (Nusantara 2024). Jakarta's case demonstrates how people around the world are already suffering from—and adapting to—a climatically changed reality and its impacts on human settlements. The Foresight Report (Government Office for Science 2011) discerned that macro factors (i.e., demographic, economic, environmental, political, social), meso factors (i.e., cost of moving, diasporic links, political/legal framework, recruitment agencies, social networks, technology), and micro factors (i.e., age, education, marital status, preferences, ethnicity, language, religion, sex, wealth) can shape decisions to migrate. In Jakarta's case, the main drivers of environmental migration have undoubtedly had to do with macro-level concern over growing exposure to hazard and impaired ecosystem services.

14.2 The Singularity of Indonesia's Capital City Relocation

Reflecting on the development and relocation of capital cities in history, Rossman (2017) observed that the chief determinants of capital city siting were geographical, military, and cultural, with some scope for politics. Rossman (2017) also made out that in the distant past some capital city relocations had been caused directly by natural cataclysmic events (e.g., an earthquake, the diversion of a river channel, or even climatic change). The relocation of Indonesia's capital city is by no means the first time that a capital city moves location: over the last 65 years, for instance, Brazil moved its capital from Rio de Janeiro to Brasília (1960); Sri Lanka moved its capital from Colombo to Sri Jayewardenepura Kotte (1982); Côte d'Ivoire moved its capital from Abidjan to Yamoussoukro (1983); Nigeria moved its capital from Lagos to Abuja (1991); Kazakhstan moved its capital from Almaty to Astana (1998); Malaysia moved its capital from Kuala Lumpur to Putrajaya (2001); Myanmar moved its capital from Yangon to Naypyidaw (2005); South Korea moved its capital from Seoul to Sejong (2007); and Tanzania finally moved its capital from Dar es Salaam to Dodoma (2023) after a 50-year wait (Schatz 2003; Wikipedia 2024c). Rachmawati et al. (2021) provided a comprehensive list of countries that relocated their capital city over the period 1946–2021. The seven aforementioned cases suggest that political, geographical, and socio-economic considerations, reminiscent of Rossman's (2017) determinants, were also the primary stimuli behind decisions to relocate the said capital cities (Ishenda and Shi 2019). According to Schatz (2003), the relocation of Kazakhstan's capital city aimed to address particularly acute nation- and state-building challenges in that post-Soviet region. However, even if natural cataclysmic events compelled a few capital city relocations in the past, the move of Indonesia's capital city from Java to eastern Borneo is the first instance of large-scale, anticipatory, and managed environmental migration in the Anthropocene. Environmental migration is "[t]he movement of persons or groups of persons who, predominantly for reasons of sudden or progressive changes in the environment that adversely affect their lives or living conditions, are forced to leave their places of habitual residence, or choose to do so, either temporarily or permanently, and who move within or outside their country of origin or habitual residence" (Sironi et al. 2019). Of course, this is not to say that a planned relocation is an easy enterprise:

> In the context of disasters or environmental degradation, including when due to the effects of climate change, [it is] a planned process in which persons or groups of persons move or are assisted to move away from their homes or place of temporary residence, are settled in

> a new location, and provided with the conditions for rebuilding their lives. (Sironi et al. 2019)

This précis casts environmental and socio-economic perspectives on the relocation of Indonesia's capital city from Jakarta to Nusantara, scheduled from 2024 to 2027; this is a topic of ever-increasing relevance and importance, subject to ongoing journalistic investigation and reporting, but about which there are very few studies to date.

14.3 Relocating Indonesia's Capital City: Key Issues

Yamoussoukro was the birthplace of Côte d'Ivoire's president: it now boasts the largest Christian cathedral in the world for a population of about 212,000 (2014 census) but still lacks the dynamism of Abidjan 40 years on. Naypyidaw features 20-lane highways with no traffic on them. Few want to walk the barren streets of Brasília (Glaeser 2012; von Reiche 2019). The major lesson learned from such capital relocations are that, past self-evident matters of adequate financing and private sector participation, moving a country's capital city calls for a sound rationale, purposeful and coordinated planning in close consultation with stakeholders, and proficient execution. Given the environmental challenges facing Jakarta, the relocation of Indonesia's capital city seems warranted but its move to eastern Borneo is a bold enterprise, ambitiously founded in eight principles: (a) designed according to nature; (b) unity in diversity; (c) connected, active, and accessible; (d) low carbon emissions; (e) circular and resilient; (f) safe and accessible; (g) security and efficiency through technology; and (h) economic opportunities for all (Nusantara 2024).

The research question that frames this précis is: "What are the environmental and socio-economic implications of relocating Indonesia's capital city, Jakarta, to Nusantara in eastern Borneo?" Each addressing a different aspect of the planned move, the two sub-questions that will help address the main question in a step-by-step manner to provide a holistic view of challenges and potential impacts are:

- How will the Government of Indonesia address the concerns raised by the relocation of Indonesia's capital city with respect to indigenous communities in Nusantara and the welfare of Jakarta's remaining population?
- How will the Government of Indonesia balance infrastructure development to create a sustainable and smart capital city in Nusantara and ensure minimal disruption to governmental operations?

14.3.1 Addressing Concerns Raised by the Relocation of Indonesia's Capital City Regarding Indigenous Communities in Nusantara and the Welfare of Jakarta's Remaining Population

The population of Indonesia's new capital city has been capped at 1.9 million residents, including 1.5 million civil servants, in consideration of environmental and land carrying capacity. The relocation of the civil servants and their family members means that 6–7 million people in toto might move to Nusantara. Therefore, the construction of Nusantara may well displace indigenous communities in East Kalimantan. Nusantara is home to 21 indigenous communities, whose elders have reportedly called for cultural preservation in Indonesia's new capital city (Fauzan 2023). The Indigenous Peoples' Alliance of Nusantara (Aliansi Masyarakat Adat Nusantara, AMAN) estimated that at least 20,000 indigenous people will be relocated as construction on Nusantara progresses (Washington and Hasibuan 2023). Tarigan and Milko (2023) reported that at least five villages with more than 100 indigenous Balik people would relocate because of the construction and that more would be uprooted. AMAN has pointed out that, irrespective of how long they and their ancestors have stayed, one of the main challenges for indigenous peoples is to prove land ownership—for lack of official certificates—so they might secure compensation in the event resettlement is forced (Washington and Hasibuan 2023).

According to Fauzan (2023) the Government of Indonesia has received support from local community leaders and provided compensation to people whose land is being used for the construction of Nusantara; further, the local government has affirmed that it is working on programs to uphold local culture. According to the Central Bureau of Statistics, as cited by the International Work Group for Indigenous Affairs (2024), the Government of Indonesia recognizes 1331 ethnic groups and an estimated number of 50–70 million indigenous peoples in the country. The Indigenous Peoples Bill that AMAN first proposed in 2009 ought to champion acknowledgment, customary land rights, empowerment, equality, justice, and protection, including the right to free, prior, and informed consent prior to development projects or business operations. AMAN presented a draft of the Indigenous Peoples Bill to the House of Representatives in 2012 but deliberations have been stalled for more than a decade (Amnesty International 2024). This is not to say there no legislation governs involuntary resettlement in Indonesia: Law No. 2 of 2012 on the Acquisition of Land for Development in the Public Interest redefined the basis on which the Government of Indonesia can procure land with clear administrative procedures and deadlines, including a legal appeal process (United Nations Environment Programme 2024). To note, above and beyond strict law and its application, indigenous peoples would also need support from genuine, impactful, and sustainable policies, programs, and initiatives. It remains to be seen how the measures related in Fauzan (2023) will mitigate the impacts of Indonesia's new capital city on the indigenous communities, for example by providing for sustainable livelihoods and employment opportunities that afford a greater level of autonomy over one's life, buttressing against the disruption of social networks, helping preserve cultural practices and traditions, and to such intents addressing the concerns of men and women on an equal footing. Where projects induce displacement and resettlement, Vanclay (2017) pinpointed concrete nonfinancial undertakings that can help reduce impoverishment risks, provide opportunity for development, and enhance shared value.

In Java, it is presupposed that the relocation of Indonesia's capital city will alleviate population density in Jakarta to decrease traffic congestion, air pollution, and river pollution, potentially improving the quality of life for the remaining residents. Jakarta, a city of 11.4 million in 2023, should remain a key urban center and continue to play a significant role in Indonesia's culture, economy, and society. Specific population projections for Jakarta after the relocation of Indonesia's capital city are not readily available. Of course, the exact impact on Jakarta's population is uncertain and will depend on such factors as future economic opportunities and infrastructure development. Not to forget, the larger Jakarta metropolitan area—known locally as Jabodetabek (an acronym of Jakarta–Bogor–Depok–Tangerang–Bekasi)—was home to over 30 million people in 2023 (World Population Review 2024); for that reason, Jakarta should still hold primary status after Indonesia's capital city relocates, which includes remaining the nation's commercial and financial center as well as its logistics hub.

14.3.2 Balancing Infrastructure Development to Create a Sustainable and Smart Capital City in Nusantara and Ensure Minimal Disruption to Governmental Operations

The world's third-largest island after Greenland and New Guinea, Borneo (about 755,000 square kilometers or 292,000 square miles) has lost half of its original forest cover to fire, illegal logging, and the expansion of palm oil and pulpwood plantations (United Nations Environment Programme 2019). Because Nusantara's population will be larger than those of Samarinda and Balikpapan—currently the biggest cities in East Kalimantan with about 1.1 million inhabitants and 688,000 inhabitants, respectively—the Government of Indonesia has formulated strategies to design Nusantara with a focus on sustainability and "smart city" principles (World Population Review 2024).

To limit the new capital city's immediate direct impacts on the environment, Nusantara is located on a coastal site about 200 kilometers (125 miles) away from the sensitive Heart of Borneo, in an area that had in earlier times been cleared for logging, coal mining, and oil palm plantations. Initiated by the World Wildlife Fund for Nature, the Heart of Borneo is a conservation program that the governments of Brunei Darussalam, Indonesia, and Malaysia launched in 2007 to protect about 220,000 square kilometers (85,000 square miles) of rich, tropical forests and their biodiversity (World Wildlife Fund for Nature 2024). Notwithstanding, it is incumbent on the Government of Indonesia to seriously and transparently conduct strategic and environmental impact assessments that integrate international best practice; a new capital city offers the opportunity to do things right. In that spirit, the Government of Indonesia is reserving three quarters of the new capital city as green, walkable areas, compared with less than 10% green space in Jakarta, which aligns with the country's target of net zero carbon emissions and 100% new and renewable energy by 2060 (Asian Development Bank 2022). Precisely, the built-up urban area of Nusantara will only make up 25% of city's total area, with 65% given over to tropical forest through reforestation of formerly cleared land and 10% ascribed to parks and food production areas: that would allow Nusantara to be a net carbon sequestration sink before 2030 and a carbon neutral city by 2045 (United Nations Economic and Social Commission for Asia and the Pacific 2023). The Nusantara Net-Zero Strategy that was launched at the United Nations Climate Conference in Dubai (COP28) on December 3, 2023 aims to achieve carbon neutrality by reversing deforestation; eliminating fossil fuel use for energy, electricity, and transportation; focusing on green and smart apartment buildings that adhere to a new model code for vertical housing; and running public transportation on electricity (BIMP-EAGA 2023).

A smart city is an urban area where information and communication technology (including the Internet of Things) help improve quality of life and the sustainability and efficiency of city operations (Gomstyn and Jonker 2023). Smart city technology will be incorporated throughout Nusantara to optimize government resources, increase sustainability, and improve the quality of life for residents in accordance with the Nusantara Smart City Blueprint, which provides strategic guidance for designing, implementing, and maintaining the smart city infrastructure (Nusantara 2024). The United States Trade and Development Agency has approved grant funding to the Nusantara Capital Authority for technical assistance in the amount of about $2.5 million to develop detailed designs, technical specifications, procurement documentation, and a capacity building strategy to support the development of Nusantara's priority smart city infrastructure (United States Trade and Development Agency 2024). The Asian Development Bank has pledged assistance: it will help design the city, assess its potential environmental and social impacts, and mobilize financing from the private and public sectors in support (Asian Development Bank 2022). Specifically, Nusantara is conceptualized as a superhub of eight clusters: clean technology industry, integrated pharmaceutical, sustainable agricultural industry, ecotourism and health, chemical and derivative products, low-carbon energy, education, and smart city and industrial (Priyambudi 2023).

To ensure minimal disruption to its operations, the Government of Indonesia is developing essential infrastructure in tune with the relocation process. The Government of Indonesia has prioritized the construction by the private sector of 47 apartment towers, located in the core area of Nusantara, for the relocation of 1.5 million civil servants and their family members from Jakarta: the construction of the towers began in August 2023 and will span 19 months, with a completion date set for December 2024 (Antara News 2023). Nusantara's infrastructure development plan also includes the construction by the private sector of a holistic living environment comprising government offices, educational facilities, greenery, sports centers, hospitals, and lakes (Nusantara 2024). Contingent on infrastructure development, the relocation of civil servants is programmed to occur in stages, with the first batch transferring from the first quarter of 2024 until 2027; the gradual transfer of civil servants to Nusantara is expected to minimize disruption to governmental operations (Nusantara 2024).

Appointed by the President of Indonesia, the Nusantara Capital City Authority oversees the relocation of Indonesia's capital city to Nusantara (Nusantara 2024). The Government of Indonesia has announced that assets in Jakarta—such as buildings owned by Indonesia's 34 ministries and the state palace—would be leased to finance the building of Nusantara (Souisa and Salim 2022). The Law on State Capital and other new regulations provide a solid basis for the relocation, ensuring that all required procedures are followed to manage the transition smoothly (Nusantara 2024). Integrated provision of physical infrastructure, information and communication technology (including the Internet of Things), transportation, and green space is in line with the best practices for capital city relocation that Rachmawati et al. (2021) made out in their literature review. The President of Indonesia is slated to celebrate the 79th anniversary of the country's declaration of independence at the site on August 17, 2024, by which time the state palace and parliament building should have been completed. The relocation of Indonesia's capital city is timetabled for completion by 2045. Specific population projections for Nusantara after the relocation of Indonesia's capital city are not readily available.

14.4 Summary, Conclusion, and Recommendation

14.4.1 Summary

The planned relocation of the seat of Indonesia's government was announced by President Widodo in 2019 and Indonesia's parliament passed a law enabling that in 2022. Home to more than 11 million people, Jakarta sits on swampy land: it has become crowded, polluted, and is sinking at an alarming rate owing to the over-extraction of groundwater. The relocation of Indonesia's capital city from Java to eastern Borneo is the first instance of large-scale, anticipatory, and managed environmental migration in the Anthropocene: the move is a momentous event with far-reaching socio-economic and environmental implications. From an environmental perspective, the relocation of Indonesia's capital city is widely seen as a chance to build a green, sustainable city from scratch. The Government of Indonesia has pledged to make Nusantara a "sustainable forest city," integrating urban development with nature. From a socio-economic perspective, the move is expected to give Jakarta much-needed breathing space, release pressure on Java, and spur economic growth in the Indonesian archipelago's east, which is among the least developed parts of the country. In the long term, the relocation of Indonesia's capital city may help realize President Widodo's aspiration to transform Indonesia into a global maritime axis at the crossroads of the Indian and Pacific oceans (Shekhar and Liow 2014).

14.4.2 Conclusion

Having through literature review characterized the motivations, challenges, and achievements of capital city relocation in Brazil (1960), Nigeria (1991), Kazakhstan (1998), Malaysia (2001), Myanmar (2005), and South Korea (2007), Syaban and Appiah-Opoku (2023) used mixed methods and triangulation of data collection to investigate the challenges and prospects facing the relocation of Indonesia's capital city. In close correspondence with the environmental challenges identified in this précis, Syaban and Appiah-Opoku (2023) established that traffic congestion, air pollution, and—especially—land subsidence (including flooding) threaten Jakarta's livability and sustainability. Consistent with this précis, Syaban and Appiah-Opoku (2023) likewise opined that the relocation of Indonesia's capital city offers a more sustainable alternative from such perspectives as modern governance, the design of innovative living systems, land availability, sustainable infrastructure, and advanced mobility and connectivity. Even so, Syaban and Appiah-Opoku (2023) conceded that high hopes for the relocation of Indonesia's capital city should be tempered by potential challenges having to do with land use changes and environmental impacts, social impacts, the securing of sufficient funding, and opposition from local community representatives (as well as sections of the national parliament, scholars from the academic realm, and individuals from the private sector).

Under President Widodo, the Government of Indonesia undertook to leave the congested, polluted, rapidly sinking, and prone to earthquakes capital city of Jakarta. The relocation of Indonesia's capital city to Nusantara affords that country a unique opportunity to reshape its future, balancing economic growth with environmental sustainability: it will have local, regional, national, and international consequences and connotes a pivotal moment in Indonesia's national trajectory. Purposeful and coordinated planning is essential to ensure that the relocation of Indonesia's capital city delivers on its promises without causing environmental damage and human harm to indigenous communities in Nusantara. At all times, the Government of Indonesia's approach must rest on four basic, interlinked, and mutually reinforcing elements of good governance: (a) accountability, (b) participation, (c) predictability, and (d) transparency. Because concerns have been raised about the potential impact on local ecosystems, it is crucial that the development of Nusantara be carried out responsibly. Because concern has also been raised about the potential impact on indigenous communities, it is likewise vital that the Government of Indonesia respect indigenous rights, conduct social impact assessments, include indigenous communities in planning and decision-making processes, provide fair compensation where relocation is necessary, and make efforts to preserve cultural heritage.

The newly established capital city of Nusantara is imbued with symbolic meaning: it has been conceptualized as "sustainable forest city" that will adhere to carbon neutrality at the geographic center of the Indonesian archipelago. Having served two five-year terms from 2014, President Widodo will step down in October 2024; ahead of that, he will have celebrated on August 17, 2024 the independence anniversary of the Republic of Indonesia in the country's new capital city of Nusantara. Prabowo Subianto, President Widodo's successor, has promised to safeguard the outgoing president's development legacy (The Economist 2024).

14.4.3 Recommendation

"Climate change is a driver of human mobility and is expected to increase the displacement of populations" (Advisory Group on Climate Change and Human Mobility 2015, p. 3). A report of the World Bank (2018) projected that climate change could force more than 140 million people to migrate within countries in Sub-Saharan Africa, South Asia, and Latin America by 2050. To note also, the global coastal population is projected to surpass one billion people by the end of this century, in a mere 75 years: as a result, sea-level rise could feature among the most costly and permanent future consequences of climate change (Hauer et al. 2019). Neumann et al. (2015) determined that, worldwide, China, India, Bangladesh, Indonesia, and Viet Nam, in descending order of magnitude, are the five countries with the largest share of population in low-elevation coastal zones, which are particularly vulnerable to hydrometeorological hazards. Other cities in Indonesia (e.g., Bima in West Nusa Tenggara, Demak Regency in Central Java, Manado in North Sulawesi, Pontianak in West Kalimantan) may in future consider environmental migration on account of climate change: over 70% of the country's population live in low-elevation coastal zones.

Clearly, planning the relocation of a city is a particularly complex undertaking. The relocation of Indonesia's capital city can serve as a beacon for other coastal cities in the country and, perhaps, for nations worldwide. The relocation of Indonesia's capital city will generate lessons aplenty and it is essential that these should not be lost, for example regarding the benefits of a systems approach to the delivery and maintenance of sustainable urban infrastructure (Serrat 2023): scholars should direct their attention toward documenting and analyzing this landmark event with continuous monitoring and evaluation that advances knowledge. The relocation of Indonesia's capital city began in earnest in 2022 but is programmed for completion in 2045; therefore, past journalistic investigation and reporting, there is considerable scope for multidisciplinary qualitative and quantitative research, both formative and evaluative, leading to new understanding in such areas as:

- **Cultural Integration in Nusantara** (e.g., studies of the cultural integration of the relocated population in Nusantara, including adoption of the local culture)
- **Cultural Preservation and Development in Jakarta and Nusantara** (e.g., research on how to preserve the cultural heritage of Jakarta and foster a cultural identity in Nusantara)

- **Environmental Impact of Nusantara** (e.g., studies of the larger environmental consequences of the relocation to Nusantara, considering factors such as land use, waste disposal, and water management, and including impacts on biodiversity, local ecosystems, and wildlife)
- **Environmental Migration Planning** (e.g., analysis of the governance mechanisms and policy decisions that underpinned and characterized the relocation of Indonesia's capital city, to serve as a case study for future large-scale, anticipatory, and managed environmental migrations)
- **Geopolitical Landscape** (e.g., research on the geopolitical implications of the relocation, both nationally and internationally, including how the Government of Indonesia positions itself in the relocation plan)
- **Governance and Public Policy** (e.g., examination of changes in governance and public policy resulting from the relocation, including potential shifts in political power and public services)
- **Health and Social Services in Nusantara** (e.g., research on the planning and provision of health and social services in Nusantara, including community centers, hospitals, and schools)
- **Impact on Jakarta** (e.g., studies of the impact on Jakarta of the migration to Nusantara, including changes in population density, housing market, job market, and public services)
- **Impact on Local Communities in Nusantara** (e.g., research could on how the relocation affects the local communities in Nusantara, including studies on changes in access to resources, cultural values, language, livelihoods, social structures, and traditions)
- **Migration Patterns in Jakarta and Nusantara** (e.g., studies of population movement patterns, including the impact on Jakarta's population and potential population growth in Nusantara)
- **Migration Responses to Sea-Level Rise** (e.g., studies of mechanisms and pathways underlying sea-level rise migration, including tipping points, thresholds and feedbacks, risk perception, and migration)
- **Socio-Economic Outcomes for the Relocated Population** (e.g., examination of the socio-economic outcomes for the relocated population, including changes in income, employment, education, family relationships, and quality of life)
- **Urban Planning and Infrastructure Development in Nusantara** (e.g., research on the planning, design, and development of Nusantara with reference to transparency and accountability, stakeholder engagement, scenario planning, data-driven decision-making, investment provisioning, international collaboration, risk mitigation, assessment of progress, etc.). Figure 14.1 provides an overview of the identified research opportunities related to the relocation of Indonesia's capital city to Eastern Borneo

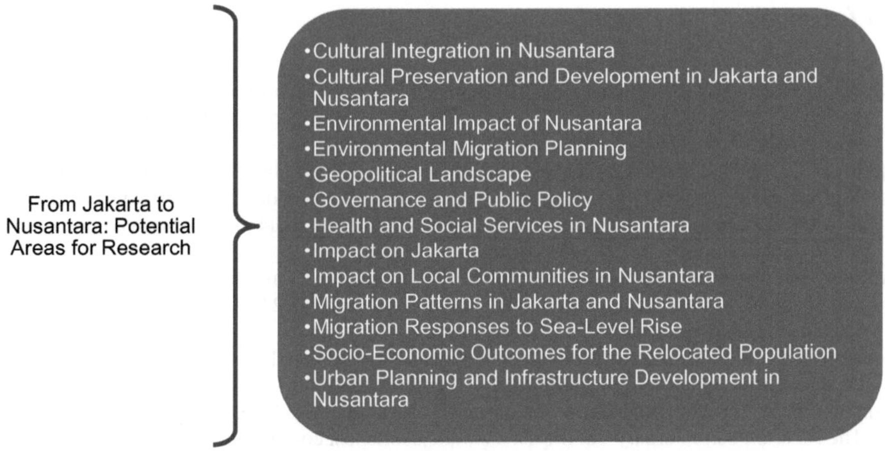

Fig. 14.1 Perspectives on the relocation of Indonesia's capital city to Eastern Borneo—potential areas for research

14.5 Postscript on Socio-Economic Dynamics

At the national level, the relocation of Indonesia's capital city implies a shift in the socio-economic dynamics of Jakarta and the rest of Indonesia: it will move the capital city closer to the nation's geographic center; it will tone down the dominance of Java and Sumatra, that in 2020 made up close to 78% of the national population, or 56% and 22%, respectively (Cabinet Secretariat of the Republic of Indonesia 2021); and, it will place the administrative center of Southeast Asia's largest economy at the center of the Brunei Darussalam–Indonesia–Malaysia–Philippines East ASEAN Growth Area, a cooperation initiative that was launched in 1994 to accelerate the socio-economic development of marginalized and geographically remote areas in these four countries (BIMP–EAGA 2024). At the regional level, the relocation of Indonesia's capital city will fortify the country's status as a dynamic Indo-Pacific power with closer proximity to areas of growing interest such as the Arafura Sea, the Celebes Sea, and—of course—the Pacific Ocean (Mishra and Wang 2022). Indonesia's new capital city will also be close to the Strait of Makassar between Borneo and Sulawesi, which is being positioned as an alternative shipping lane to the very busy Strait of Malacca: the strategic, maritime connectivity of the move cannot be overstated (Mishra and Wang 2022). At the local level, the transfer of Indonesia's capital city is expected to stimulate economic growth in East Kalimantan: the influx of civil servants and the development of infrastructure should create spillover effects in the form of new business opportunities and jobs and boost economic prosperity in the province (Mishra and Wang 2022).

References

Aditya A, Sipahutar T, Rahadiana R (2019) Indonesia picks Borneo for new capital amid Jakarta gridlock. Bloomberg, August 26. https://www.bloomberg.com/news/articles/2019-08-26/jokowi-picks-borneo-for-new-capital-as-jakarta-nears-gridlock

Advisory Group on Climate Change and Human Mobility (2015) Human mobility in the context of climate change—Recommendations from the advisory group on climate change and mobility COP 20 Lima, Peru. https://www.iom.int/sites/g/files/tmzbdl486/files/migrated_files/pbn/docs/Human-Mobility-in-the-context-of-Climate-Change.pdf

Amnesty International (2024) Indonesia: submission to the UN Committee on Economic, Social, and Cultural Rights: 75th session, 12 February–1 March 2024, January 11. https://www.amnesty.org/en/documents/asa21/7572/2024/en/

Antara News (2023) Ministry commences building 47 flat towers in Nusantara, August 28. https://en.antaranews.com/news/292200/ministry-commences-building-47-flat-towers-in-nusantara

Aqil A (2023) Turkey earthquake a lesson and a warning for Indonesia: meteorology agency. The Jakarta Post, February 26. https://www.thejakartapost.com/Indonesia/2023/02/26/turkey-earthquake-a-lesson-and-a-warning-for-Indonesia-bmkg.html

Asian Development Bank (2022) Building a green and smart city in Indonesia's new capital, April 11. https://seads.adb.org/news/building-green-and-smart-city-Indonesias-new-capital

Associated Press (2022) Indonesia's capital is rapidly sinking into the sea, January 26. https://www.npr.org/2022/01/26/1075720551/jakarta-Indonesia-sinking-into-java-sea-new-capital

BIMP–EAGA (2023) Nusantara net-zero strategy launched at COP28 in Dubai, December 11. https://bimp-eaga.asia/article/nusantara-net-zero-strategy-launched-cop28-dubai

BIMP–EAGA (2024). https://www.bimp-eaga.asia/

Brittanica (2024a) Jakarta. In Brittanica.com. https://www.britannica.com/place/Jakarta

Brittanica (2024b) Dutch East Indies. Brittanica.com. https://www.britannica.com/place/Dutch-East-Indies

Brittanica (2024c) Ring of fire. Brittanica.com. https://www.britannica.com/place/Ring-of-Fire

Cabinet Secretariat of the Republic of Indonesia (2021) Statistics Indonesia releases 2020 census results, January 23. https://setkab.go.id/en/statistics-Indonesia-releases-2020-census-results/

Cabinet Secretariat of the Republic of Indonesia (2022) Ratified law on capital marks new capital development, February 21. https://setkab.go.id/en/ratified-law-on-capital-marks-new-capital-development/

Carrubba A (2020) Rotten river: life on one of the world's most polluted waterways—Photo essay. Guardian, November 2. https://www.theguardian.com/global-development/2020/nov/02/rotten-river-life-on-one-of-the-worlds-most-polluted-waterways-photo-essay

Earth5R (2020) Jakarta environmental problems and circular economy solutions, July 29. https://earth5r.org/jakarta-circular-economy/

Economist (2024) Prabowo Subianto will be the next president of Indonesia, April 26. https://www.economist.com/interactive/2024-Indonesia-election-tracker

Fauzan M (2023) Elders call for Indigenous cultural preservation in new Indonesia capital. Mongabay, July 27. https://news.mongabay.com/2023/07/elders-call-for-indigenous-cultural-preservation-in-new-Indonesia-capital/

Gambolati G, Teatini P (2021) Land subsidence and its mitigation. The Groundwater Project. https://books.gw-project.org/land-subsidence-and-its-mitigation/

Glaeser E (2012) Brasília is a warning to urban dreamers. Financial Times, December 9. https://www.ft.com/content/52a967d0-409d-11e2-8f90-00144feabdc0

Gokkon B (2022) Three-fourths of waste in Jakarta's notoriously polluted rivers is plastic. Mongabay, August 29. https://news.mongabay.com/2022/08/three-fourths-of-waste-in-jakartas-notoriously-polluted-rivers-is-plastic/

Gomstyn A, Jonker A (2023) What is a smart city? IBM, November 22. https://www.ibm.com/topics/smart-city

Government Office for Science (2011) Migration and global environmental change: future challenges and opportunities. https://www.gov.uk/government/publications/migration-and-global-environmental-change-future-challenges-and-opportunities

Hauer M, Fussell E, Mueller V, Burkett M, Call M, Abel K, McLeman R, Wrathall D (2019) Sea-level rise and human migration. Na Rev Earth Environ 1(1):28–39

Hutton G (2013) Cleaning up one of the world's most polluted places. World Bank Blogs, December 19. https://blogs.worldbank.org/water/cleaning-one-world-s-most-polluted-places

International Work Group for Indigenous Affairs (2024) Indigenous peoples in Indonesia. https://www.iwgia.org/en/Indonesia/4224-iw-2021-Indonesia.html

Ishenda D, Shi G (2019) Determinants in relocation of capital cities. J Public Admin Gov 9(4):200–220

Jakarta Post (2018) Jakarta's river water severely polluted: Bappenas, February 2. https://www.thejakartapost.com/news/2018/02/02/jakartas-river-water-severely-polluted-bappenas.html

Lin M, Hidayat R (2018) Jakarta, the fastest-sinking city in the world. British Broadcasting Corporation, August 12. https://www.bbc.com/news/world-asia-44636934

Mishra R, Wang P (2022) From Jakarta to Nusantara: making sense of Indonesia's capital shift. Lowy Institute, October 31. https://www.lowyinstitute.org/the-interpreter/jakarta-nusantara-making-sense-Indonesia-s-capital-shift

NASA Earth Observatory (2019) As Jakarta grows, so do the water issues, September 11. https://earthobservatory.nasa.gov/images/148303/as-jakarta-grows-so-do-the-water-issues

National Aeronautics and Space Administration (2024) What is climate change? https://climate.nasa.gov/what-is-climate-change/

Neumann B, Vafeidis A, Zimmermann J, Nicholls R (2015) Future coastal population growth and exposure to sea- level rise and coastal flooding—A global assessment. PLOS ONE 10(6):1–34

Nichols W (2021) Asian cities in eye of environmental storm—global ranking. Environmental Risk Outlook 2021. Verisk Maplecroft, May 12. https://www.maplecroft.com/insights/analysis/asian-cities-in-eye-of-environmental-storm-global-ranking/

Nusantara (2024) Toward the Indonesia 2045 vision: a global city for all. https://www.ikn.go.id/en

Priyambudi I (2023) Nusantara an energizing site for construction industry. The Jakarta Post, January 23. https://www.thejakartapost.com/paper/2023/01/23/nusantara-an-energizing-site-for-construction-industry.html

Rachmawati R, Haryono E, Ghiffari R, Reinhart H, Permatasari F, Rohmah A (2021) Best practices of capital city relocation in various countries: literature review. E3S Web of Conferences 325:1–7. https://www.e3s-conferences.org/articles/e3sconf/abs/2021/101/e3sconf_icst2021_07004/e3sconf_icst2021_07004.html

von Reiche L (2019) Moving capital cities can often be a costly folly. Financial Times, May 3. https://www.ft.com/content/e212eaa0-6ccd-11e9-80c7-60ee53e6681d

Renaldi A (2022) Indonesia's giant capital city is sinking. Can the government's plan save it? National Geographic, August 1. https://www.nationalgeographic.co.uk/science-and-technology/2022/08/Indonesias-giant-capital-city-is-sinking-can-the-governments-plan-save-it

Rossman V (2017) Capital cities: varieties and patterns of development and relocation. Routledge

Schatz E (2003) When capital cities move: the political economy of nation and state building. Working Paper No. 303. Helen Kellogg Institute for International Studies. https://kellogg.nd.edu/documents/1596

Serrat O (2023) Localizing climate action in Latin American and Caribbean cities. Unpublished manuscript, Georgetown University. https://www.researchgate.net/publication/376032182_Localizing_Climate_Action_in_Latin_American_and_Caribbean_Cities

Serrat O (2024) Understanding environmental migration: a start-up kit. Unpublished manuscript, Georgetown University. https://www.researchgate.net/publication/380178533_Understanding_Environmental_Migration_A_Start-Up_Kit

Shekhar V, Liow J (2014) Indonesia as a maritime power: Jokowi's vision, strategies, and obstacles ahead. Brookings Institution, November 7. https://www.brookings.edu/articles/Indonesia-as-a-maritime-power-jokowis-vision-strategies-and-obstacles-ahead/

Sironi A, Bauloz C, Emmanuel M (Eds) (2019) Glossary on migration. International Migration Law, No. 34. International Organization for Migration. https://publications.iom.int/books/international-migration-law-ndeg34-glossary-migration
Souisa H, Salim N (2022) Nusantara is set to be the new capital of Indonesia, but what will happen to Jakarta? ABC News, January 29. https://www.abc.net.au/news/2022-01-30/what-will-happen-to-jakarta-when-Indonesia-builds-a-new-capital/100784566.
Syaban A, Appiah-Opoku S (2023) Building Indonesia's new capital city: an in-depth analysis of prospects and challenges from current capital city of Jakarta to Kalimantan. Urban Plann Transp Res 11(1):1–29. https://www.tandfonline.com/doi/full/10.1080/21650020.2023.2276415
Tarigan E, Milko V (2023) Why Indonesia is moving its capital from Jakarta to Borneo. PBS Newshour, March 9. https://www.pbs.org/newshour/world/why-Indonesia-is-moving-its-capital-from-jakarta-to-borneo
United Nations Economic and Social Commission for Asia and the Pacific (2023) Navigating the challenges of new city development for Nusantara, Indonesia's future capital, September 18. https://www.unescap.org/blog/navigating-challenges-new-city-development-nusantara-Indonesias-future-capital
United Nations Environment Programme (2019) Deforestation in Borneo is slowing, but regulation remains key, February 18. https://www.unep.org/news-and-stories/story/deforestation-borneo-slowing-regulation-remains-key
United Nations Environment Programme (2024) Law on the acquisition of land for development in the public interest (No. 2 of 2012). https://leap.unep.org/en/countries/id/national-legislation/law-acquisition-land-development-public-interest-no-2-2012
United States Trade and Development Agency (2024) United States announces new support for sustainable development of Nusantara, March 7. https://id.usembassy.gov/united-states-announces-new-support-for-sustainable-development-of-nusantara/
Van Mead N (2016) The world's worst traffic: can Jakarta find an alternative to the car? Guardian, November 23. https://www.theguardian.com/cities/2016/nov/23/world-worst-traffic-jakarta-alternative
Vanclay F (2017) Project-induced displacement and resettlement: from impoverishment risks to an opportunity for development? Impact Assess Proj Appraisal 35(1):3–21. https://www.tandfonline.com/doi/full/10.1080/14615517.2017.1278671
Washington J, Hasibuan S (2023) "Like we don't exist": indigenous fear Indonesia new capital plan. Al Jazeera, March 15. https://www.aljazeera.com/news/2023/3/15/like-we-dont-exist-indigenous-fear-Indonesia-new-capital-plan
Wikipedia (2024a) History of Indonesia. Wikipedia.org. https://en.wikipedia.org/wiki/History_of_Indonesia
Wikipedia (2024b) Floods in Jakarta. Wikipedia.org. https://en.wikipedia.org/wiki/Floods_in_Jakarta
Wikipedia (2024c) List of former national capitals. Wikipedia.org. https://en.wikipedia.org/wiki/List_of_former_national_capitals
World Bank (2018) Groundswell: preparing for internal climate migration. https://openknowledge.worldbank.org/entities/publication/2be91c76-d023-5809-9c94-d41b71c25635
World Health Organization (2024) Air quality, energy, and health. https://www.who.int/teams/environment-climate-change-and-health/air-quality-and-health/health-impacts/types-of-pollutants
World Population Review (2024). https://worldpopulationreview.com/
World Wildlife Fund for Nature (2024) Heart of Borneo. https://wwf.panda.org/discover/knowledge_hub/where_we_work/borneo_forests/

Chapter 15
Understanding Environmental Migration: A Start-Up Kit

Abstract Environmental migrants are people who are compelled to leave their home region or choose to do so, either temporarily or permanently, predominantly because of slow- or rapid-onset changes in their environment that compromise their well-being or livelihood. Climate change is a driver of human mobility that is expected to compound the displacement of populations: developing more effective responses to this global challenge calls for improvements in definitions, theories, normative frameworks, and research.
Human mobility has always been influenced by environmental factors. However, the severity and scope of environmental shocks have intensified because of climate change, resulting in increased internal and international migration and displacement. As our planet grapples with mounting environmental challenges, it is important to understand environmental migration to develop more effective responses to this global challenge.

15.1 Re: Environmental Migration Patterns

Environmental migrants are people who are compelled to leave their home region or choose to do so, either temporarily or permanently, predominantly as a consequence of slow- or rapid-onset changes in their environment that compromise their well-being or livelihood (International Organization for Migration 2024a). Environmental migration has been classified as voluntary, involuntary, or planned. Today, the spatial and temporal patterns of environmental migration that one can discern include internal and international migration (with links between the two), permanent migration, seasonal migration, circular migration, and evacuations. Migration across international borders has for long been sensationalized in media: presumably for that reason, international environmental migration has been the subject of research; however, small-scale and localized environmental displacement, for example in consequence of flood events, is a much more common occurrence. One can also distinguish patterns of sending and receiving countries that form interconnected systems. Patterns of environmental migration will vary between countries

O. Serrat, *Anthropogenic Solutions for Climate Change*,
https://doi.org/10.1007/978-981-96-8654-4_15

depending on the ways in which the environment is being affected by slow-onset events (e.g., desertification, drought, erosion, sea-level rise) or rapid-onset events (e.g., flash floods; cyclones; storms; natural hazards such as earthquakes and volcanic eruptions). Much of sub-Saharan Africa, for instance, is increasingly subject to drought and desertification while the Pacific Islands are especially susceptible to sea-level rise. There will be variations within affected communities too: the less affluent and the more vulnerable (e.g., the elderly) will be impacted first; the better-off will weather the changes a little better. The unique consequences that will result in places that are affected by environmental change have to do with the loss of livelihoods, which will likely lead to migration at some point in the future. Livelihoods hang on environmental, financial, human, infrastructure, and social capital (Serrat 2017a). Loss of environmental capital due to climate change will impact the other forms of capital and lead to general impoverishment.

15.2 Re: Environmental Migration Theories

Theories of environmental migration have drawn insights from the broader field of international migration. Core international migration theories include: (a) neoclassical economics, which pays attention to wage differences, income maximization, and the cost of migration; (b) dual labor market, which investigates the demand for migrant labor to fill low-wage sectors; (c) world systems, which sees that globalization links core and periphery countries, creating migration flows, (d) the new economics of labor migration, which emphasizes household-level decision-making and how remittances can mitigate social and environmental stressors; (e) network theory, which posits that social ties reduce migration costs and perpetuate flows, potentially beyond the control of governments; (f) institutional theory, which holds that organizations facilitate migration, thus making it self-sustaining; (g) cumulative causation, which advances that migration acts cumulatively to change the economic and social landscape in a way that perpetuates itself; and (h) migration systems, which studies patterns of sending and receiving countries that form interconnected systems. Theories are wont to promote new explanations but in most instances rest on previous explanations: thus, for example, neoclassical economics and the new economics of labor migration both place an accent on income maximization and remittances. That said, explanations of environmental migration must perforce be more complex owing to the multicausal nature of the phenomenon: to wit, it is not just the environmental driver or natural hazard that calls for attention: rather, it is the interactions of cultural, demographic, economic, political, security, and social factors that lead people to migrate or to remain in place, conditioned also by macro-, micro-, and meso-level factors. For these reasons, neglecting to consider the natural environment engenders models that attribute migration solely to economic or social factors while failure to incorporate migration theory in climate scenarios begets unsophisticated assessments.

15.3 Re: Normative Frameworks for Environmental Migration

Environmental migration challenges global governance: it does not fit in existing normative international and national frameworks, even if it is a deeply human issue that affects individuals and families in profound ways. Today, normative frameworks only recognize refugees (who flee persecution, cannot return home due to fear, and are protected by international law); migrants (who move for various reasons, are not protected by international refugee law, and can return home if they choose); and internally displaced persons (who are displaced within their own country, are not refugees under international law, but are protected by human rights law). However, environmental migrants are not recognized by any international or national normative framework. To boot, both at international and national levels, different organizations are tasked with regulating refugees, migrants, and internally displaced persons. At the international level, the Convention Relating to the Status of Refugees (1951) is the key legal document that defines who is a refugee and outlines the rights and responsibilities of states; it is internationally binding (United Nations High Commissioner for Refugees 2024a). On the other hand, the Global Compact on Refugees (2018) is a nonbinding agreement that outlines a comprehensive approach to addressing refugee situations (United Nations High Commissioner for Refugees 2024b). And, the Global Compact for Safe, Orderly and Regular Migration (2018) is another nonbinding agreement that broadly frames global cooperation on migration issues (Office of the United Nations High Commissioner for Human Rights 2024). However, none of the aforementioned normative frameworks makes a case for environmental migrants or environmental refugees. It cannot be said that the practices associated with each of the three normative frameworks mentioned conflict with or support one another: they were formulated with different objectives in mind, none of which addresses the special circumstances of environmental migration. Possibly, the Convention Relating to the Status of Refugees (1951) could be amended to include a special provision for environmental refugees. There have also been calls for environmental migration to be associated with human rights, and thus fall under the Universal Declaration of Human Rights (1948): however, this is not yet on the cards owing to widespread international diffidence (United Nations 2024). Country governments have grappled with the multifaceted implications of environmental displacement and need to accelerate the development of policies to address it in the face of misleading terms (e.g., "climate refugees," "environmental refugees") and data challenges. To note, the "hard laws" that archetypally govern the movement of people across international borders provide enforceability and legal certainty but cannot cater to the nuances and complexities of movements that are driven by climate change. As an alternative, "soft laws," viz., declarations, guidelines, nonbinding legal norms, and principles, are more likely to quickly fill gaps, provide guidance, and encourage state practice; they are also easier to negotiate and amend as needed.

15.4 Re: Environmental Migration Research

Environmental migration is a field in the process of formation: it draws on many disciplines (e.g., anthropology, demography, economics, international law, politics, sociology) with human and physical geography making particular contributions (Serrat 2024). Research must be multidisciplinary: failure to adopt a pluridisciplinary approach to the examination of environmental migration can only constrain the value of research and limits its explanatory power or proposed solutions, be that regarding mitigation, adaptation, climate resilience, environmental risk, or environmental exposures. Research on environmental migration must also answer for the multicausal nature of environmental migration and displacement. And, for greater relevance, research on environmental migration should move beyond a narrow focus on aggregate numbers of people who move because of climate change to also investigate, say, the importance of place attachment; the resilience of communities; efforts to remain in situ; and the phenomenon of adaptive migration as temporary, seasonal, and circular. Notwithstanding the addition of collection tools from the International Organization for Migration (2024b), measuring mobility in the context of environmental and climate change is not easy owing to the miscellaneousness of data, be it sourced from population registers, census, surveillance systems, or surveys (i.e., cross-sectional, multiple cross-sectional, longitudinal). The immobility of trapped populations is rarely the subject of research. What is more, problems of definition aggravate the difficulty that research on environmental migration faces. Refugees and internally displaced persons are "better" subjects of research: they are more narrowly defined and accounted for under international and national normative frameworks. In contrast, environmental migrants are yet to be given the status they must have if their special case is to be addressed and treated: this has implications for research because it makes the environmental migrant invisible. Not to forget, the term "migration" does not capture the diverse ways in which people can become mobile over relatively short distances in response to slow- or rapid-onset changes in their environment; for that reason, there have been calls for more research to examine what is referred to as "climate mobility." Beyond the securitization narrative and agenda, considerations of climate mobility do not assume that the movement of people is always monocausal or unidirectional, nor do they label it as intrinsically positive or negative. Today, estimates of the number of environmental migrants vary greatly: the International Organization for Migration reported that forecasts range from 25 million to 1 billion by 2050, including people who move either within their countries or across borders on a permanent or temporary basis, with 200 million being the "most widely cited" estimate (International Organization for Migration 2023). To help develop more effective responses, researchers must generate better findings (Pellini and Serrat 2017; Serrat 2017b): they need to know what to call people who move for environmental reasons; how to characterize the complicated nexus of climate change, sudden or gradual changes in the environment, and conflict; and how to design policies, programs, and projects for environmental migrants, both internal and international.

References

International Organization for Migration (2023) A complex nexus: what are the estimates? https://www.iom.int/complex-nexus

International Organization for Migration (2024a) Environmental migration. https://environmental-migration.iom.int/environmental-migration

International Organization for Migration (2024b) Information management tools. https://emergencymanual.iom.int/information-management-tools

Office of the United Nations High Commissioner for Human Rights (2024) Global compact for safe, orderly, and regular migration. https://www.ohchr.org/en/migration/global-compact-safe-orderly-and-regular-migration-gcm

Pellini A, Serrat O (2017) Enriching policy with research. In: Knowledge solutions: tools, methods, and approaches to drive organizational performance. Springer, pp 937–944

Serrat O (2017a) The sustainable livelihoods approach. In: Knowledge solutions: tools, methods, and approaches to drive organizational performance. Springer, pp 21–26

Serrat O (2017b) Linking research to practice. In: Knowledge solutions: tools, methods, and approaches to drive organizational performance. Springer, pp 3–6

Serrat O (2024) Understanding the environment in Africa: a start-up kit. Unpublished manuscript, Georgetown University. https://www.researchgate.net/publication/377806396_Understanding_the_Environment_in_Africa_A_Start-Up_Kit

United Nations (2024) Universal declaration of human rights. https://www.un.org/en/about-us/universal-declaration-of-human-rights

United Nations High Commissioner for Refugees (2024a) The 1951 refugee convention. https://www.unhcr.org/about-unhcr/who-we-are/1951-refugee-convention

United Nations High Commissioner for Refugees (2024b) The global compact on refugees. https://www.unhcr.org/about-unhcr/who-we-are/global-compact-refugees

Part VI
Environmental Peacebuilding

Chapter 16
Caring for Our Common Home: Ecocritical Perspectives on Pope Francis's *Laudato Si'*

Abstract Pope Francis was a pivotal voice in the global fight against the ecological devastation being wrought on the Earth. His groundbreaking encyclical letter, *Laudato Si'*, remains influential, pressing for a profound ecological conversion, urging humanity to take responsibility for the planet's health and emphasizing the moral imperative to protect our common home. This précis provides an ecocritical analysis of *Laudato Si'*, highlighting its significance in contemporary environmental discourse.

In a papacy marked by historic milestones, Pope Francis was a pivotal voice in the global fight against the ecological devastation being wrought on the Earth. His groundbreaking encyclical letter, *Laudato Si'* (Francis 2015), presses for a profound ecological conversion, urging humanity to take responsibility for the planet's health and emphasizing the moral imperative to protect "our common home" (Francis 2015, para. 1). The pastoral letter is distinguished by its exceptional global reach and influence, its incorporation of scientific findings, and its holistic approach that uniquely addresses the interconnectedness of cultural, economic, environmental, and social issues. This précis provides an ecocritical analysis of *Laudato Si'*, highlighting its significance in contemporary environmental discourse. Ultimately, the pontifical document serves as both a call to action and a guiding light for sustainable practices worldwide.

16.1 The Roman Catholic Church's Green Revolution

The Roman Catholic Church's stance on ecological concerns has evolved significantly over time (United States Conference of Catholic Bishops 2001). Traditionally, its teachings focused on the stewardship of creation, underscoring humanity's responsibility to care for the Earth as God's creation. This perspective is rooted in biblical texts, such as the Book of Genesis, which describes humanity's role in tending to the Garden of Eden.

O. Serrat, *Anthropogenic Solutions for Climate Change*,
https://doi.org/10.1007/978-981-96-8654-4_16

In recent decades, the Church has acknowledged the pressing need to reduce environmental degradation. This shift demonstrates the Church's ability to adapt and respond to global challenges. In 1979, Pope John Paul II proclaimed Saint Francis of Assisi (1181/82–1226), an Italian mystic, poet, and Catholic friar who founded the Franciscan Order, the heavenly patron of those who promote ecology. Saint Francis was celebrated for his connection with nature and his intense respect for all creation, often referring to the sun and moon as "brother" and "sister" (Britannica 2024). Notably, in the *Common Declaration on Environmental Ethics* (John Paul II and Bartholomew I 2002), Pope John Paul II (1920–2005) placed an accent on the moral dimensions of ecological concerns, calling for a global ecological conversion, or change of heart. Pope Benedict XVI (1927–2022), often referred to as the "Green Pope," underlined the importance of environmental stewardship and sustainable development in his encyclical *Caritas in Veritate* (Benedict XVI 2009) and his 2010 World Day of Peace message, *If You Want to Cultivate Peace, Protect Creation* (Benedict XVI 2010).

Born Jorge Mario Bergoglio on December 17, 1936, in Buenos Aires, Argentina, to Italian immigrant parents, Pope Francis was the 266th Pope of the Church. He was ordained a priest in the Society of Jesus in 1969 and became a bishop in 1992. In 1998, he was appointed Archbishop of Buenos Aires and was named a cardinal in 2001 (Francis and Ivereigh 2020). Elected Bishop of Rome on March 13, 2013, he was the first Jesuit pope, the first pope from the Americas, and the first non-European pope in over a millennium. Tellingly, Pope Francis chose his papal name in honor of Saint Francis of Assisi.

Under the new pontiff, the Church intensified its advocacy for environmental stewardship and sustainable development, urging comprehensive and urgent action for environmental protection. This advocacy confronts critical issues such as climate change, deforestation, desertification, habitat destruction, loss of biodiversity, invasive species, ocean acidification, overfishing, plastic waste, pollution, resource depletion, and soil degradation. Shortly after his election, Pope Francis initiated the preparation of *Laudato Si'*. He enlisted climate and environmental science experts to gather the most accurate data on the state of the planet. He then invited theologians to engage with this data in a global dialogue with experts. Theologians and scientists collaborated until they reached a synthesis. *Laudato Si'* firmly positioned ecology as a central theme in Church discourse (Deane-Drummond 2017). Strategically timed for publication just before the 2015 United Nations Climate Change Conference (COP21) in Paris, many attribute the historic undertaking of the so-called Paris Agreement (United Nations 2015) to the influence of this work and the efforts of Pope Francis. The Paris Agreement aspires to keep the global temperature rise this century well below 2°C above pre-industrial levels and pursue efforts to limit the increase to 1.5°C.

The pontiff aimed for *Laudato Si'* to reach a worldwide audience and spark international discussion. The Church remains one of the largest religious institutions globally, with a substantial presence in various countries and cultures. Its reach and influence can mobilize millions of people towards sustainable practices and policies. Currently, there are approximately 1.4 billion baptized Catholics worldwide, reflecting a slight rise from previous years, with notable growth in Africa and Asia (Quiñones 2024). Besides, while secular societies may not adhere to religious doctrines, the ethical and moral arguments presented by religious leaders can still engage a large audience.

16.2 *Laudato Si'*: A Call for Ecological Conversion

"All it takes is one good person to restore hope!" Pope Francis exclaimed in *Laudato Si'* (Francis 2015, para. 71). During the pontiff's papacy, the world faced significant challenges, including pandemics; rising economic inequalities; political polarization; migration and refugee crises; religious persecution and violence in various parts of the world (e.g., China, India, Middle East, Myanmar, Nigeria, Pakistan); technological advancements and ethical concerns; and conflicts in Syria (2011–), Yemen (2014–), Myanmar (2021–), Ukraine (2022–), and Israel-Palestine (2023–). These challenges shaped Pope Francis's papacy and his efforts to address them through advocacy, dialogue, and pastoral care. Moreover, the urgent necessity for climate action has never been more critical. Understandably, the pope's tenure was marked by a strong accent on social justice, advocacy for the marginalized, and environmental stewardship, reflecting his commitment to tackling these critical global issues.

In *Laudato Si'*, Pope Francis reflected on the world's ecological concerns and did not mince words: "Due to an ill-considered exploitation of nature, humanity runs the risk of destroying it and becoming in turn a victim of this degradation" (Francis 2015, para. 4). "Every effort to protect and improve our world entails substantial changes in lifestyles, models of production and consumption, and the established structures of power which today govern societies" (Francis 2015, para. 5). "The [E]arth, our home, is beginning to look more and more like an immense pile of filth" (Francis 2015, para. 21). "[N]owadays we must forcefully reject the notion that our being created in God's image and given dominion over the [E]arth justifies absolute domination over other creatures" (Francis 2015, para. 67). "A global consensus is essential for confronting the deeper problems, which cannot be resolved by unilateral actions on the part of individual countries" (Francis 2015, para. 164). As the first papal letter devoted to environmentalism, *Laudato Si'* urges readers to fundamentally reconsider our relationship with the planet. Praised by educators, environmentalists, scientists, and leaders of various faiths for its balanced poise, the pastoral letter continues to significantly influence international environmental discourse, shaping the Church's environmental policies and encouraging sustainable practices worldwide (Handley 2016).

In support of its call to action for environmental stewardship and social justice, *Laudato Si'* articulated several key themes:

- **Biodiversity.** The encyclical makes obvious the vital significance of preserving biodiversity and recognizing the intrinsic value of all living creatures.

- **Climate Change.** The encyclical points to the broad scientific agreement on climate change and its catastrophic impacts, especially on the poor and vulnerable.
- **Consumerism and Waste.** The encyclical critiques the pervasive culture of consumerism and the throwaway mentality, advocating for sustainable and responsible consumption patterns.
- **Ethical and Spiritual Dimensions.** The encyclical calls for a deep moral and spiritual renewal, urging individuals and communities to adopt a more respectful and caring attitude towards nature.
- **Global Solidarity.** The encyclical highlights the necessity of global cooperation and solidarity to tackle environmental challenges, stressing the importance of dialogue and collective action.
- **Integral Ecology.** The encyclical points to the interconnectedness of all creation, promoting an approach that integrates environmental, economic, social, and cultural dimensions because "Everything is related" (Francis 2015, para. 92). As stated in the encyclical, the elements of integral ecology that enable us to consider every aspect of the global crisis are environmental, economic, and social ecology; cultural ecology; ecology of daily life; the principle of the common good; and justice between generations.
- **Social Justice.** The encyclical links environmental degradation to social injustice, exposing how the poor and vulnerable, including marginalized and indigenous peoples, are disproportionately affected and campaigning for their rights and dignity.
- **Technological and Economic Factors.** The encyclical considers the role of technology and economic systems in environmental degradation, calling for a critical evaluation to ensure they align with sustainable and ethical principles.

Several other religious texts have emphasized environmental stewardship and ethical responsibility. The *Earth Charter*, an international declaration endorsed by various religious groups, recognizes the interconnectedness of all life and the need for sustainable development and social justice (Earth Charter Initiative 2000). Islamic teachings on the unity of creation and the responsibility of humans to act as stewards of the Earth, known as *tawhid*, are rooted in the *Quran* and *Hadith* (Nasr 1996). *The Green Bible* is a special edition of the Bible that highlights scriptures related to environmental stewardship, including essays and a green index to guide readers in understanding the biblical basis for caring for creation (Green Bible 2008). Ecumenical Patriarch Bartholomew I, often referred to as the "Green Patriarch," has issued numerous encyclicals and statements on the ethical and spiritual dimensions of ecological responsibility (Bartholomew I 2003). Hindu texts like the *Bhagavad Gita* and the *Vedas* point out the sanctity of nature and the responsibility of humans to safeguard and maintain the environment (Dwivedi 1993). Similarly, the concept of interdependence and the principle of non-harming (*ahimsa*) in Buddhism encourage a respectful and mindful relationship with the natural world (Kabilsingh 1998). However, *Laudato Si'* stands out for its unparalleled global reach and influence, its integration of scientific findings, and its comprehensive approach that uniquely attends to the interconnectedness of cultural, economic, environmental, and social issues.

Building on the themes of *Laudato Si'*, Pope Francis advocated for urgent action against climate change, lamenting its disproportionate impact on the poor and vulnerable. *Our Mother Earth* (Francis 2020) set forth a Christian vision of ecology, gathered from his writings and discourses. In the exclusive new essay that concludes *Our Mother Earth* (Francis 2020), Pope Francis developed a "theology of ecology" in a most ethical and spiritual discourse. This last chapter shares thoughts on how a Christian vision of care for our common home transcends a purely secular approach to ecology. The pope remarked that humanity's capacity for communion conditions the state of creation, suggesting that it is humanity's destiny to determine the destiny of the universe. This theological perspective underscores the interconnectedness of all creation and seeks a global ecological conversion, urging individuals, communities, and societies to adopt sustainable practices and recognize the moral and ethical dimensions of environmental stewardship. Also in 2020, the pontiff introduced the Laudato Si' Action Platform (https://laudatosiactionplatform.org/), a seven-year initiative promoting sustainability across various sectors. The platform suggests concrete steps and resources for individuals and communities to implement sustainable practices. His apostolic exhortation, *Laudate Deum* (Francis 2023), built on *Laudato Si'*, summoning urgent and coordinated climate action. Pope Francis's environmental advocacy extended to international forums, urging nations to honor the Paris Agreement (United Nations 2015) and mitigate climate change.

16.3 Ecocriticism: Bridging Literature and Environmental Discourse

Ecocriticism is an interdisciplinary field that encourages a deeper appreciation of the natural world and promotes environmental awareness and activism through literary analysis. In the face of escalating environmental crises, ecocriticism has become increasingly significant in contemporary environmental discourse (Bergthaller et al. 2014). By integrating insights from various fields, ecocriticism contributes to a comprehensive understanding of ecological problems and solutions (Glotfelty and Fromm 1996). Specifically, it observes how literature portrays nature, environmental issues, and human interactions with the natural world. Narratives shape societal values and behaviors; by analyzing texts, ecocritics seek to understand how literature reflects, critiques, or influences environmental attitudes and policies.

As environmental crises grow more urgent, ecocriticism has gained significant traction in contemporary environmental discourse (Garrard 2014). Garrard (2023) identified six overarching political and philosophical frames that understand environmental crises in their own way: cornucopia, ecofeminism, ecological modernization, new materialism, political ecology and environmental justice, and radical ecology. Key concepts reflecting the diversity within ecocriticism include animals, the Earth, indigeneity, pastoral, pollution, and wilderness (Garrard 2023). A central tenet of ecocriticism is the recognition of the interconnectedness between humans and the environment. This perspective challenges the traditional anthropocentric view that places humans at the center of the universe, advocating instead for a more rounded understanding of the world (Glotfelty and Fromm 1996).

Pastoral poets who foreshadowed contemporary eco-activism include William Wordsworth (1770–1850), Gerard Manley Hopkins (1844–1889), and G. K. Chesterton (1874–1936) (Bate 1991). Notable early works of fiction and non-fiction in the genres of nature writing and transcendentalism, as well as environmental science and advocacy, include *Walden; or, Life in the Woods* (Thoreau 1854) and *Silent Spring* (Carson 1962). In contemporary environmental discourse, works of fiction, nonfiction, and poetry that treat themes such as climate change, deforestation, and pollution serve as powerful tools for environmental advocacy. By engaging readers emotionally and intellectually, these texts foster a deeper connection to the natural world and a greater sense of responsibility for its preservation (Buell 2005).

Moreover, ecocriticism underscores the importance of diverse perspectives in understanding environmental issues. It recognizes that different cultures and communities have unique relationships with the environment, shaped by their histories, traditions, and values. By examining a wide range of literary works from various cultural contexts, ecocritics uncover alternative ways of thinking about and relating to nature. An inclusive approach is necessary to deal with contemporary environmental challenges, as it fosters cross-cultural dialogue and collaboration (Clark 2011).

Another significant aspect of ecocriticism is its focus on the ethical implications of environmental issues. It asserts the moral responsibilities of individuals, societies, and governments in preventing environmental crises. By analyzing how literature portrays the consequences of environmental degradation, ecocritics shed light on the ethical dilemmas and injustices associated with these issues. For example, many literary works draw attention to the disproportionate impact of environmental problems on marginalized communities, highlighting the need for environmental justice (Rueckert 1978).

In addition to its contributions to literary studies, ecocriticism has influenced other disciplines, such as philosophy, political science, and sociology. Its interdisciplinary nature allows for a more comprehensive understanding of environmental issues, integrating varied insights into complex ecological problems. This approach remains particularly relevant in contemporary environmental discourse, where the interconnectedness of ecological, economic, and social systems is increasingly recognized (Garrard 2014; Heise 2008).

Furthermore, ecocriticism has stimulated new forms of creative expression that engage with environmental themes. Eco-documentaries, eco-fiction, eco-films, eco-photography, eco-poetry, and environmental memoirs are just a few examples of genres that have emerged in response to the growing awareness of ecological issues

(Bryson 2002; Dwyer 2010; Kimmerer 2013; Parini 1999; Schneider-Mayerson 2018; Williams 2017). These works not only contribute to the literary landscape but also serve as important cultural artifacts that reflect and shape our collective understanding of the environment (Glotfelty and Fromm 1996).

16.4 *Laudato Si'* Through the Lens of Ecocriticism

Key themes of ecocriticism that are particularly relevant to *Laudato Si'* are:

- **Critique of Technocratic and Industrial Paradigms.** Ecocriticism critiques the dominance of technocratic and industrial paradigms that prioritize economic growth over environmental sustainability. The Vatican directive challenged the technocratic and industrial mindset and promoted a more sustainable and ethical approach to development.
- **Cultural Dimensions.** Ecocriticism appraises the cultural dimensions of human–nature relationships. The Vatican directive appreciated the spiritual significance of nature and sponsored a cultural shift towards greater environmental awareness and stewardship.
- **Environmental Justice.** Ecocriticism often brings up issues of environmental justice, focusing on how environmental degradation disproportionately affects marginalized communities. The Vatican directive urged social justice and equitable distribution of resources, reasoning there was a moral imperative to protect the poor and vulnerable.
- **Ethical Responsibility.** Ecocriticism promotes the idea of ethical responsibility towards the environment. The Vatican directive called for a global ecological conversion, urging individuals, communities, and societies to adopt sustainable lifestyles and practices.
- **Integration of Science and Humanities.** Ecocriticism often integrates scientific knowledge with literary and cultural analysis to illuminate environmental issues. The Vatican directive incorporated scientific evidence on climate change and environmental degradation in favor of an interdisciplinary approach to solving ecological problems.
- **Interconnectedness.** Ecocriticism grasps the interconnectedness of all living beings and the environment. The Vatican directive clarified that environmental, social, economic, and political issues were deeply intertwined and stressed the need for a holistic approach to environmental issues.

These themes illustrate how *Laudato Si'* aligned with and contributed to the broader discourse of ecocriticism, offering a comprehensive and ethically grounded perspective on environmental issues. In addition, *Laudato Si'* drew attention to:

- **Consumerism.** The apostolic letter considered the culture of consumerism and irresponsible development and recommended sustainable practices and a shift towards simpler, more mindful living.

- **Dialogue and Solidarity.** The apostolic letter encouraged dialogue and solidarity, promoting collaboration among scientists, policymakers, religious leaders, and all sectors of society to develop practical solutions to environmental challenges.
- **Ecological Conversion.** The apostolic letter pressed for a transformation in how humanity views and interacts with the environment, urging everyone to take responsibility for the planet's health.
- **Ethics and Spirituality.** The apostolic letter framed environmental stewardship as a core component of the Christian faith, reiterating the moral imperative to protect our common home.
- **Impact on the Poor.** The apostolic letter pointed out the disproportionate impact of environmental degradation on the poor and vulnerable, calling for environmental justice and equitable solutions.
- **Integral Ecology.** The apostolic letter introduced the concept of integral ecology, promoting a holistic approach to environmental crises because ecology encompasses far more than just the care of nature.

Laudato Si' was a comprehensive document that took up many key themes of ecocriticism. However, there are a few themes, some of which were brought to light by Garrard (2023), that were not as prominently featured or were approached differently:

- **Animal Studies.** Animal studies scrutinize the ethical considerations of human–animal relationships and the treatment of animals. While the pontifical document acknowledged the importance of caring for all creatures, it did not thoroughly appraise the ethical dimensions of human–animal relationships.
- **Apocalypse.** Apocalyptic ecocriticism often critiques apocalyptic narratives and sponsors proactive environmental stewardship over fatalistic outlooks. Similarly, the pontifical document aimed to inspire hope and action for environmental stewardship, promoting global dialogue and collective effort rather than yielding to apocalyptic fears or indifference.
- **Deep Ecology.** Deep ecology champions a shift in human consciousness, recognizing the intrinsic value of all living beings beyond their utility to humans. While the pontifical document highlighted the interconnectedness of all life and the need for a holistic approach, it primarily focused on the ethical and moral responsibilities of humans rather than advocating for a radical shift in consciousness.
- **Ecofeminism.** Ecofeminism examines the links between the exploitation of women and the environment, arguing that both are rooted in patriarchal structures. While the pontifical document strongly advocated for social justice and gender equality, it did not explicitly frame these issues within the context of ecofeminism.
- **Postcolonial Ecocriticism.** Postcolonial ecocriticism looks into the intersections of environmental issues and colonial histories, focusing on how colonialism has impacted both the environment and the poor and vulnerable. While the pontifical document explicitly rebuked the exploitation of the poor and vulnerable, it did

not delve into the specific historical and ongoing impacts of colonialism on the environment.

- **Urban Ecocriticism.** Urban ecocriticism analyzes the relationship between urban environments and ecological issues, focusing on how cities can be designed and managed to be more sustainable. The pontifical document discussed the importance of sustainable development and the need for ecological conversion, but it did not specifically touch on the particular concerns of urban ecocriticism.

These themes highlight areas where *Laudato Si'* could have focused its discussion to align more closely with the conventional discourse of ecocriticism, if such a need had arisen. Notwithstanding, the encyclical remains a most significant and influential document that, with remarkable brevity, attended to many critical environmental and social issues in under 180 concise pages (246 paragraphs).

16.5 Intersections of Faith and Philosophy: *Laudato Si'* and Morton's Ecological Thought

Timothy Morton is a contemporary philosopher and ecocritic known for his work on ecology, literature, and object-oriented ontology, a new school of thought in metaphysics that explores the existence and agency of nonhuman objects. Object-oriented ontology challenges the traditional privileging of human existence over nonhuman entities and posits that objects exist independently of human perception and interaction. Consequently, it rejects the idea that human existence should prevail over the existence of nonhuman objects. Morton has authored numerous books, including *The Ecological Thought* (Morton 2010), *Hyperobjects: Philosophy and Ecology after the End of the World* (Morton 2013), and *Dark Ecology: For a Logic of Future Coexistence* (Morton 2016). His work often probes the intersections of ecological studies and object-oriented thought, advocating for a radical rethinking of human relationships with nonhuman entities and the environment.

Embarking on a journey through *Laudato Si'* with the insights of Morton (2010) is akin to opening a dialogue between two discerning perspectives on our relationship with the environment. Morton (2010) dismantles the illusion of nature as a distant, untouched realm, urging us to see the interconnected web of life that includes the artificial and the synthetic. *Laudato Si'*, on the other hand, advocated for an integral ecology that weaves together the cultural, economic, environmental, and social threads of our existence. In distinct ways but with the same intention, these works invite us to reconsider our place in the world and embrace a holistic approach to ecological stewardship.

Despite their different origins—one rooted in religious doctrine and the other in environmental philosophy—*Laudato Si'* and Morton (2010) showed up the urgent need for a shift in how humanity perceives and interacts with the environment. These works converged on the idea of interconnectedness:

- **Critique of Consumerism and Technological Impact.** Both Morton (2010) and *Laudato Si'* critiqued the prevailing culture of consumerism and the detrimental impact of technology on the environment. "Modern economic structures have drastically affected the environment" (Morton 2010, p. 4). "Do we have to go into outer space to care for Earth? Do we need high technology? Do we need Google Earth to imagine Earth? Is Western science and power the only path to ecological awareness?" (Morton 2010, p. 25). Similarly, *Laudato Si'* critiqued the throwaway culture and the unsustainable consumption patterns that contribute to environmental degradation (Handley 2016). "[W]e cannot fail to consider the effects of environmental deterioration, current models of development and the throwaway culture on people's lives" (Francis 2015, para. 43). The encyclical invited a critical evaluation of technological advancements and economic practices to ensure they align with sustainable and ethical principles.
- **Ethical and Spiritual Dimensions.** While Morton's (2010) work was grounded in environmental philosophy, *Laudato Si'* incorporated strong ethical and spiritual dimensions. "Cultural limitations in different eras often affected the perception of […] ethical and spiritual treasures, yet by constantly returning to their sources, religions will be better equipped to respond to today's needs" (Francis 2015, para. 200). The encyclical encouraged a spiritual awakening and a moral response to the ecological crisis, urging individuals and communities to adopt a more respectful and caring attitude towards nature (Handley 2016). Morton's (2010) ecological thought, though not explicitly spiritual, also implied an ethical responsibility towards the environment, emphasizing the need for a shift in how we think about and engage with the world around us. "The ecological thought must imagine economic change: otherwise it's just another piece on the board game of capitalist ideology?" (Morton 2010, p. 19).
- **Global Solidarity and Collective Action.** Morton (2010) and *Laudato Si'* ascribed much importance to global solidarity and collective action in the face of environmental challenges. *Laudato Si'* encouraged global cooperation and dialogue to tackle environmental issues, stressing the need for a unified response (Iheka 2017). "We require a new and universal solidarity" (Francis 2015, para. 14). Morton (2010), too, highlights the necessity of thinking beyond localism and embracing a global perspective on ecological interconnectedness. "If we think the ecological thought, two things happen. Our perspective becomes very vast. More and more aspects of the Universe become included in the ecological thought. At the same time, our view becomes very profound" (Morton 2010, p. 38). Morton (2010) and *Laudato Si'* recognized that addressing the ecological crisis requires a concerted effort from all sectors of society. "Perhaps the sentiment we're going for is not 'We can because we must,' but rather 'We must because we are'" (Morton 2010, p. 124).
- **Interconnectedness and Integral Ecology.** At the heart of both Morton (2010) and *Laudato Si'* was the concept of interconnectedness. Morton (2010) challenged the notion of nature as a separate, pristine entity, arguing instead for an ecological thought that acknowledges the entanglement of all life forms, including the synthetic and the artificial. "The ecological thought imagines interconnectedness,

which I call the mesh. Who or what is interconnected with what or with whom? The mesh of interconnected things is vast, perhaps immeasurably so. Each entity in the mesh looks strange. Nothing exists all by itself, and so nothing is fully 'itself.' There is curiously 'less' of the Universe at the same time, and for the same reasons, as we see 'more' of it" (Morton 2010, p. 15). Morton (2010) argued that all forms of life are entangled in a vast, intricate mesh, where no being or object exists independently from the ecological web. This idea resonates with *Laudato Si*'s concept of integral ecology (Handley 2016; Iheka 2017). "To seek only a technical remedy to each environmental problem which comes up is to separate what is in reality interconnected and to mask the true and deepest problems of the global system" (Francis 2015, para. 111). Comparably, Morton (2010) and *Laudato Si'* advocated for a comprehensive understanding of ecology that transcends traditional boundaries and recognizes the interdependence of all life forms.

Morton (2010) and *Laudato Si'* offered complementary perspectives on the ecological crisis. While Morton's (2010) work provided a philosophical framework for understanding ecological interconnectedness, *Laudato Si'* integrated this understanding with a call for ethical and spiritual transformation. In unison, Morton (2010) and *Laudato Si'* underscored the urgent need for a holistic approach to environmental stewardship, one that recognizes the interconnectedness of all life and the moral imperative to protect our common home.

Laudato Si' drew on a wide range of sources, including previous papal teachings, scientific research, and theological reflections. While Morton (2010) is noteworthy in the field of ecocriticism and shares some thematic similarities with *Laudato Si'*, such as the interconnectedness of all life and the critique of anthropocentrism, there is no direct evidence to suggest that Morton (2010) specifically influenced the apostolic letter.

16.6 *Laudato Si'*: A Catalyst for Ecological Transformation

Laudato Si' demonstrated that religious leaders can play a crucial role in environmental discourse by leveraging their moral authority and social influence to shift attitudes and behaviors, foster a sense of stewardship among their followers, and inspire action. This aligns with theories of moral authority and social influence. Their unique position enables them to bridge the gap between scientific understanding and ethical imperatives, making complex environmental issues more accessible and urgent to a broader audience. By framing environmental care as an ethical and spiritual duty, religious leaders can galvanize their followers to adopt more sustainable practices and advocate for policies that protect the planet.

Laudato Si' was a landmark document in the realm of environmental advocacy. It highlighted the interconnectedness of all creation and the moral imperative to care for our common home. By framing environmental issues through the lens of Catholic social teaching, Pope Francis enriched the conversation to include ethical and spiritual dimensions, making it clear that environmental degradation was not just a scientific or economic issue, but a moral crisis. He challenged humankind to rethink its relationship with nature and adopt more sustainable lifestyles.

Laudato Si' set a precedent for future environmental advocacy by stressing the importance of integrating ethical and spiritual perspectives into the discourse. It encouraged other religious leaders and institutions to take a more active role in promoting environmental stewardship. The encyclical also stressed the need for global cooperation and solidarity, suggesting that effective environmental action requires a collective effort that transcends national and cultural boundaries. This approach can hearten a more inclusive and comprehensive strategy for combating environmental challenges. Potential areas for further research include:

- **Case Studies of Successful Religious Environmental Movements.** Document and analyze successful environmental movements led by religious groups, identifying key factors that contributed to their success and lessons that can be applied to other contexts.
- **Ethical Frameworks for Environmental Decision-Making.** Develop and analyze ethical frameworks that incorporate religious principles to guide environmental decision-making and policy development.
- **Impact of Religious Teachings on Environmental Behavior.** Study the influence of religious teachings on the environmental attitudes and behaviors of adherents, identifying effective strategies for promoting sustainable practices within faith communities.
- **Interfaith Environmental Initiatives.** Investigate how different religious traditions can collaborate on environmental projects and advocacy, fostering a united front for ecological sustainability.
- **Role of Religious Institutions in Policy Advocacy.** Reflect on how religious institutions can engage in policy advocacy at local, national, and international levels to support environmental initiatives.

These areas of research can offer valuable insights into the intersection of religion and environmentalism, helping to shape more inclusive approaches to the ecological crisis. "Start by doing what is necessary, then what is possible, and suddenly you are doing the impossible," Saint Francis of Assisi (n.d.) is famously quoted as saying. One might reasonably assume that Pope Francis was in full agreement.

16.7 *Laudato Si'*: Sparking New Directions in Ecocritical Research

Laudato Si' has significant implications for ecocriticism. Here are some key points:

- **Call for Ecological Conversion.** The concept of "ecological conversion" in the pastoral letter insisted on a transformation in how humans relate to nature. This idea can motivate ecocritics to weigh up themes of transformation and renewal in environmental literature.

- **Critique of Technocratic Paradigms.** The pastoral letter challenged the technocratic paradigm that prioritizes technological and economic growth over environmental sustainability. This critique resonates with ecocritical analyses that challenge dominant narratives of progress and development.
- **Holistic Approach to Environmental Issues.** The pastoral letter advocated for a holistic approach to environmental problems, considering social, economic, and cultural factors. This aligns with ecocriticism's interdisciplinary nature, encouraging scholars to explore the interconnectedness of various aspects of human life and the environment.
- **Integration of Ethical and Spiritual Dimensions.** The pastoral letter brought a moral and spiritual perspective to environmental issues, underlining the ethical responsibility of humans to care for the Earth. This enriches ecocriticism by adding a layer of ethical and spiritual analysis to environmental literature.
- **Promotion of Global Solidarity.** The pastoral letter drove home the need for global cooperation and solidarity to address environmental challenges. This perspective can embolden ecocritics to consider global and cross-cultural dimensions in their analyses.

These implications underscore the potential of *Laudato Si'* to enrich and expand ecocriticism. By offering new research avenues, the encyclical addressed Buell's (2005) challenge for the discipline to establish its significance beyond academia.

16.8 *Ad Astra per Aspera*

This précis deliberated several themes, starting with the Church's intensified focus on environmental ethics, highlighted by Pope Francis's momentous encyclical, *Laudato Si'*. It then transitioned to ecocriticism, analyzing how the central themes of this interdisciplinary field were reflected in the Vatican directive. Next, by focusing on a specific work of ecocriticism, the précis examined the intersections of faith and philosophy as seen in *Laudato Si'* and Morton (2010). Great works inherently provide ample food for thought. Consequently, the précis also considered how *Laudato Si'* might serve as a catalyst for further religious environmental movements and posited that it could enhance and enlarge the field of ecocriticism itself.

In the prologue to *Let Us Dream: The Path to a Better Future* (Francis and Ivereigh 2020), the pontiff reflected on a line in Friedrich Hölderlin's hymn, *Patmos*: "But where the danger is, also grows the saving power" (cited in Francis and Ivereigh 2020, p. 6), meaning that even in the midst of crisis, there is always a way to escape destruction. In *Laudato Si'*, Pope Francis echoed this sentiment, offering a beacon of hope and a call to action for humanity. "Humanity still has the ability to work together in building our common home" (Francis 2015, para. 13). Pope Francis described *Laudato Si'* as a social encyclical, explaining that "[t]he green and the social go hand in hand" (Francis and Ivereigh 2020, p. 32).

In the final chapters of *Laudato Si'*, Pope Francis outlined transformative pathways for dialogue and action to break free from the cycle of self-destruction that threatens our world. These pathways include: fostering international economic dialogue on environmental issues; crafting innovative national and local policies; ensuring transparency in decision-making processes; harmonizing politics and economy for human well-being; and bridging the gap between religions and science (Francis 2015, paras. 163–201; Serrat 2024d). Pope Francis also championed ecological education and spirituality "to enable the development of new convictions, attitudes, and forms of life" (Francis 2015, para. 202).

Epilogue

Sight-Worn

Sight-worn lidless eyes turning inward
Seek darkness in the past,
Lie me in a shaded recess of the mind
Where kind memories diffuse a softer glow.
I mold words for your ears.
My fingers trace the contours of your thoughts,
A touch ever so light the pattern arches back,
Murmuring worlds, tasting the feel.
A groomed engine then,
We become an instant and breathing must be willed.
It is colder now.
The wire leading your voice through the maze underground chills it hard metallic—
I shall wear a tie for the rest of my life.

In this my Tokyo train (another rush hour day)
The shuttling of matter in entrails to the tune
Of a perched employee orchestrating the boom.
Toranomon. Mutes roam the corridors head bowed,
Trying the step to ground level, and emerge in the rain.
A hundred little gray worlds bloom in a flurry of umbrellas,
Hop graceless o'er puddles and puncture in doorways.
Sowing raincoats, myriads rush to the masque
Donning smiles on the way.

My pen as a lance probes and pokes at the wall till,
Self-propelled and sudden-sucked,
Spinning and swirling through the envelope,
Dizzying free, a part becomes the whole.
I am world, my hair foliage.
A courser wind-swept, my blood is new sap wailing.
A crumpled page.

In this his unbuilt house,
My father shall cajole the jet lag out our bones
And with it cement stones that to the hill shall draw the family in pairs.
Amber-suffused limbs by the green apple fire,
Bristling up down its beams of warmth to keep,
We shall suffer the dark to signal the way home.
The glasses at our hands—chalices rapt in ice—bitter sweetness encompass.
A gentian-flavored drink sings another for the road,
Tomorrow being Monday.

Please leave me not,
Through want of prevenient grace,
To partake with dodoes, dinosaurs, aborigines too,
Of a stuporous fate.
'Fore the last yards of a once endless race,
Run now with the faltering office gait of
I that stitched, pant, and feverish spit more froth in unseeing eyes,
Lord of Life, show me Thy Grace. (Serrat 2023)

References

Bartholomew I (2003) Encyclicals and statements on environmental issues. Ecumenical Patriarchate.

Bate J (1991) Romantic ecology: Wordsworth and the environmental tradition. Routledge

Benedict XVI (2009) Caritas in Veritate: on integral human development in charity and truth. Vatican Press. https://www.vatican.va/content/benedict-xvi/en/encyclicals/documents/hf_ben-xvi_enc_20090629_caritas-in-veritate.html

Benedict XVI (2010) If you want to cultivate peace, protect creation, January 1. https://www.vatican.va/content/benedict-xvi/en/messages/peace/documents/hf_ben-xvi_mes_20091208_xliii-world-day-peace.html

Bergthaller H, Emmett R, Johns-Putra A, Kneitz A, Lidström S, McCorristine S, Pérez Ramos I, Phillips D, Rigby K, Robin L (2014) Mapping common ground: ecocriticism, environmental history, and the environmental humanities. Environ Humanit 5(1):261–276

Britannica (2024) St. Francis of Assisi. https://www.britannica.com/biography/Saint-Francis-of-Assisi

Bryson JS (ed) (2002) Ecopoetry: a critical introduction. University of Utah Press

Buell L (2005) The future of environmental criticism: environmental crisis and literary imagination. Wiley-Blackwell

Carson R (1962) Silent spring. Houghton Mifflin

Clark T (2011) The Cambridge introduction to literature and the environment. Cambridge University Press

Deane-Drummond, C. (2017). A primer in ecotheology. Cascade Companions.

Dwivedi O (1993) Hinduism and ecology. In: Chapple C, Tucker ME (eds) Worldviews and ecology. Orbis Books, pp 193–208

Dwyer J (2010) Where the wild books are: a field guide to ecofiction. University of Nevada Press

Earth Charter Initiative (2000) The Earth charter. Earth Charter Commission

Francis P (2015) Laudato Si': on care for our common home. Vatican Press. https://www.vatican.va/content/francesco/en/encyclicals/documents/papa-francesco_20150524_enciclica-laudato-si.html

Francis P (2020) Our mother Earth: a Christian reading of the challenge of the environment. Our Sunday Visitor

Francis P (2023) Laudate Deum: on the urgency of climate action. Vatican Press. https://www.vatican.va/content/francesco/en/apost_exhortations/documents/20231004-laudate-deum.html

Francis P, Ivereigh A (2020) Let us dream: the path to a better future. Simon & Schuster

Garrard G (2014) The Oxford handbook of ecocriticism. Oxford University Press

Garrard G (2023) Ecocriticism, 3rd edn. Routledge

Glotfelty C, Fromm H (1996) The ecocriticism reader: landmarks in literary ecology. University of Georgia Press

Green Bible (2008) The Green Bible. HarperOne

Handley G (2016) Laudato Si' and the postsecularism of the environmental humanities. Environ Humani 8(2):277–284

Heise U (2008) Sense of place and sense of planet: the environmental imagination of the global. Oxford University Press

Iheka C (2017) Pope Francis' integral ecology and environmentalism for the poor. Environ Ethics 39(3):243–259

John Paul II, Bartholomew I (2002) Common declaration on environmental ethics. Vatican Press. https://www.vatican.va/content/john-paul-ii/en/speeches/2002/june/documents/hf_jp-ii_spe_20020610_venice-declaration.html

Kabilsingh C (1998) Buddhism and ecology: the interconnection of dharma and deeds. Motilal Banarsidass

Kimmerer, R. W. (2013). Braiding sweetgrass: indigenous wisdom, scientific knowledge, and the teachings of plants. Milkweed Editions.

Morton T (2010) The ecological thought. Harvard University Press

Morton T (2013) Hyperobjects: philosophy and ecology after the end of the world. University of Minnesota Press

Morton T (2016) Dark ecology: for a logic of future coexistence. Columbia University Press

Nasr SH (1996) Religion and the order of nature. Oxford University Press

Parini J (1999) Poems for a small planet: contemporary American nature poetry. Middlebury College Press

Quiñones K (2024) The Catholic Church by the numbers: more Catholics but fewer vocations. Catholic News Agency, April 8. https://www.catholicnewsagency.com/news/257316/the-catholic-church-by-the-numbers-more-catholics-but-fewer-vocations

Rueckert W (1978) Literature and ecology: an experiment in ecocriticism. In: Glotfelty C, Fromm H (eds) The ecocriticism reader: landmarks in literary ecology. University of Georgia Press, pp 105–123

Schneider-Mayerson M (2018) The influence of climate fiction: an empirical survey of readers. Environ Humanit 10(2):473–500

Serrat O (2023) Mere words, these: 32 poems in free verse. Troubador Publishing

Serrat O (2024a) ICYMI: climate security rhymes with national security. Unpublished manuscript, Georgetown University. https://www.researchgate.net/publication/378545941_ICYMI_Climate_Security_Rhymes_with_National_Security

Serrat O (2024b) Beggar thy neighbor to beggar thyself: why China and the United States must cooperate to address climate change. Unpublished manuscript, Georgetown University. https://www.researchgate.net/publication/380212272_Beggar_Thy_Neighbor_to_Beggar_Thyself_Why_China_and_the_United_States_Must_Cooperate_to_Address_Climate_Change

Serrat O (2024c) Redefining norms: a modern-day Marshall Plan for climate adaptation and climate justice. Unpublished manuscript, Georgetown University. https://www.researchgate.net/publication/380530326_Redefining_Norms_A_Modern-Day_Marshall_Plan_for_Climate_Adaptation_and_Climate_Justice

Serrat O (2024d) Bridging worlds: comparative epistemology and the mind–body debate. Unpublished manuscript, Georgetown University. https://www.researchgate.net/publication/380611280_Bridging_Worlds_Comparative_Epistemology_and_the_Mind-Body_Problem

Thoreau H (1854) Walden; or, life in the woods. Ticknor and Fields

United Nations (2015) Paris Agreement. https://unfccc.int/sites/default/files/english_paris_agreement.pdf

United States Conference of Catholic Bishops (2001) Global climate change: a plea for dialogue, prudence, and the common good, June 15. https://www.usccb.org/resources/global-climate-change-plea-dialogue-prudence-and-common-good

Williams F (2017) The nature fix: why nature makes us happier, healthier, and more creative. W.W. Norton & Company

Chapter 17
Disciplinary Perspectives on Environmental Peacebuilding: Insights from Environmental Science and International Relations

Abstract By integrating environmental considerations into peacebuilding initiatives, we can create lasting solutions that address recurring root causes of conflict, such as competition over natural resources, environmental degradation, and resource scarcity. Toward this, we must adopt a holistic and cross-disciplinary approach, embracing diverse viewpoints and integrating multiple areas of expertise. This précis delves into how insights from environmental science and international relations might specifically contribute to environmental peacebuilding.

17.1 Harnessing Nature for Peace: The Power of Environmental Peacebuilding

Envision a world where nature's elements—air, minerals, soil, water, and other resources—serve as the keystones of peace (Conca and Dabelko 2002). This transformative approach, environmental peacebuilding, addresses shared environmental challenges to promote cooperation, trust, and sustainable development (United Nations Environment Programme 2023).

By integrating environmental considerations into peacebuilding initiatives, we can create lasting solutions that target the root causes of conflict, such as competition over natural resources, environmental degradation, and resource scarcity (Homer-Dixon 1999). Examples of environmental peacebuilding initiatives include: climate change adaptation projects, collaborative water management, community-based natural resource management, cross-border wildlife conservation, forest conservation and management, joint environmental monitoring, marine conservation agreements, reforestation and land restoration projects, renewable energy projects, sustainable agriculture programs, transboundary air pollution agreements, and transboundary protected areas. This holistic approach addresses the intertwined challenges of environmental sustainability and social stability, ensuring that communities can thrive with sustainable livelihoods and a healthy environment (Serrat 2017). By fostering cooperation, trust, and resilience, we can lay the foundation for enduring peace and long-term stability (Dresse et al. 2019; Heffington et al. 2019a;

O. Serrat, *Anthropogenic Solutions for Climate Change*,
https://doi.org/10.1007/978-981-96-8654-4_17

Ide 2021; Krampe et al. 2024; Scoones 1998; United Nations Environment Programme 2023).

Unavoidably, effective environmental peacebuilding initiatives require inclusive stakeholder engagement, robust collaboration, innovative strategies, continuous adaptation, and thorough evaluation of lessons learned (Scoones 1998; Serrat 2024a). Each initiative should thoroughly assess the strategic situation, define the desired outcomes, develop the necessary means to achieve these outcomes, design ways to leverage these means effectively, and assess the associated costs and risks (Heffington et al. 2019b).

Above all, to achieve success, environmental peacebuilding must adopt a holistic and cross-disciplinary approach, embracing diverse viewpoints and integrating multiple areas of expertise (Sommer and Fassbender 2024). While specialized knowledge is valuable, relying solely on the reductionism that single disciplines encourage can oversimplify complex problems by analyzing them in isolation. Anchored in positivism, pragmatism, and realism, and employing reductionism when necessary, the deepest insights often emerge through abduction, deduction, and induction at the crossroads of various disciplines, with pragmatism often proving particularly effective (Slingerland 2008). In environmental peacebuilding, as in other areas in general, single disciplines should complement each other rather than compete for supremacy. Environmental peacebuilding is inherently integrative, drawing insights from both the natural and social sciences to understand and manage the intersections of environmental challenges and conflict dynamics, and to create solutions that are both sustainable and transformative. For instance, enhanced academic rigor suggests the combined efforts of environmental science and international relations, as examples discussed in this précis, can significantly upgrade the effectiveness of environmental peacebuilding.

17.2 Research Question

According to the Global Atlas of Environmental Justice (https://ejatlas.org/), there were more than 4200 documented environmental conflicts worldwide as of early 2024.[1] Given the global significance of environmental conflicts and the critical roles that both environmental science and international relations can play in understanding and addressing these issues, this précis delves into how insights from these two

[1] The Global Atlas of Environmental Justice (https://ejatlas.org/) classifies environmental conflicts across ten categories: biodiversity conservation conflicts, biomass and land conflicts (forests, agriculture, fisheries and livestock management), fossil fuels and climate justice/energy, infrastructure and built environment, industrial and utilities conflicts, mineral ores and building materials extraction, nuclear, tourism recreation, waste management, and water management. The Global Atlas of Environmental Justice (https://ejatlas.org/) is a work in progress: newly documented cases and information are continuously added. However, many conflicts remain undocumented and new ones arise.

disciplines might specifically contribute to environmental peacebuilding.[2] Environmental science provides crucial knowledge about natural systems and sustainable management practices, while international relations offer insights into the economic, political, and social dynamics of peace and conflict.[3] By combining these perspectives, we can develop more comprehensive and effective strategies for environmental peacebuilding.

This précis delves into the question: How can environmental science and international relations contribute to environmental peacebuilding initiatives? It explores how these disciplines address issues from different perspectives, uniquely illuminate the research question, provide answers, and diverge in their investigative approaches. Finally, it suggests ways to integrate these perspectives to make progress on the research question.

17.3 Quid Environmental Peacebuilding?

17.3.1 Emergence and Evolution of Environmental Peacebuilding

Environmental peacebuilding emerged in the 1980s, recognizing that competition over natural resources, environmental degradation, and resource scarcity can exacerbate social tensions and contribute to conflict (Homer-Dixon 1999). This concept is particularly relevant in today's interconnected world, where climate change, deforestation, water scarcity, and other environmental challenges threaten local and

[2] Environmental peacebuilding initiatives are not solely international. While many conflicts involve shared resources that cross national borders—such as air quality, forests, groundwater aquifers, marine ecosystems, mountain ranges, rivers and lakes, and wildlife migration corridors—peacebuilding can also occur within a single country. Local factors driving peacebuilding include climate change, cultural and historical significance, economic development, policy effectiveness, environmental degradation, religious and ethnic tensions, resource scarcity, and social tensions (Peace Direct & Alliance for Peacebuilding 2020). This précis focuses on transboundary peacebuilding initiatives to examine the specific role of international relations.

[3] International relations, a subfield of political science, addresses issues such as conflict, diplomacy, global economic relations, global governance, human rights, international law, and war (Mingst and Arreguín-Toft 2016). Political science, the broader field, explores the theory and practice of politics and government at all levels—local, national, and international—and includes other subfields such as comparative politics, political theory, public administration, and public policy (Roskin and Cord 2019). This précis highlights the relevance of international relations to environmental peacebuilding due to its emphasis on interactions between countries and global challenges that transcend national borders (Goldstein and Pevehouse 2017). These challenges encompass environmental issues, conflict and cooperation, international agreements and treaties, the role of international organizations, and diplomacy and foreign policy. In summary, political science provides the broader context of governance and policy, while international relations offers tools and frameworks to address complex, transboundary environmental challenges and the collaborative efforts required for peacebuilding.

global stability (Heffington et al. 2019b). For example, overfishing in coastal regions disrupts livelihoods, leading to economic hardship and social unrest. Deforestation in the Amazon affects local social unrest, biodiversity, and global climate change, impacting ecological balance. Climate change impacts are global, causing displacement, food insecurity, and heightened tensions. These examples underscore the urgent need for environmental peacebuilding as a comprehensive approach to addressing the multifaceted challenges of environmental conflicts at both local and global levels (Dresse et al. 2018).

17.3.2 *The Complex Interplay Between Environmental Factors and Conflict Dynamics*

Initially, environmental peacebuilding focused on understanding the impact of environmental factors on conflict dynamics. The field has since evolved to address the complex interplay between environmental factors and conflict dynamics, incorporating strategies for leveraging environmental management to promote peace (Krampe et al. 2024; Krampe and Swain 2024).

To address the multifaceted challenges of environmental conflicts, environmental peacebuilding must integrate insights from multiple disciplines, requiring an understanding of cultural, ecological, economic, political, and social dimensions.[4] This cross-disciplinary approach fosters a comprehensive understanding of environmental factors and conflict dynamics. By harnessing the strengths of multiple disciplines, environmental peacebuilding can create a more robust framework for addressing the environmental challenges that contribute to conflict while fostering lasting peace and stability. Figure 17.1 identifies the core disciplines that contribute to environmental peacebuilding, emphasizing the interconnected nature of ecological sustainability and conflict resolution.

---

[4] For example, initiatives aimed at reducing deforestation in the Amazon often face resistance due to conflicting conservation and economic interests (e.g., logging) that bypass legislation. Collaborative efforts involving indigenous leaders, ecologists, economists, and sociologists using real-time satellite monitoring and stricter penalties for violations could create sustainable strategies that protect ecosystems. In Colombia, post-conflict efforts to restore degraded lands and integrate former combatants into conservation work face logistical and social challenges. Environmental scientists, social workers, and conflict resolution specialists could energize these initiatives and boost their long-term success. In the Lake Chad region, environmental degradation and water scarcity have worsened conflicts among communities reliant on the lake, which spans four countries (Serrat 2023). A cross-disciplinary approach involving diplomats, climate scientists, agricultural experts, and conflict resolution specialists could address both environmental and social aspects, offering a path to resilience. In the Mekong River basin, dam construction has fueled international disputes over water rights, ecological damage, and impacts on fisheries and agriculture (Serrat 2024b). Cross-disciplinary teams of hydrologists, environmental scientists, diplomats, and local representatives could strengthen transboundary water governance. In every scenario, stronger political will could turn challenges into opportunities.

Fig. 17.1 Principal disciplines relevant to environmental peacebuilding. (*Note*. Other relevant disciplines may include environmental education, environmental engineering, environmental ethics, environmental psychology, geology, hydrology, public health, sociology, and urban planning, depending on the context)

17.3.3 Core Goals of Environmental Peacebuilding

Necessarily, environmental peacebuilding initiatives must be tailored to address specific local environmental, political, and social contexts. Therefore, these initiatives aim to tackle diverse challenges through various goals (Krampe and Swain 2024). Drawing on Matthew et al. (2009), Conca and Dabelko (2002), and Conca and Wallace (2009), it is possible to identify core goals that environmental peacebuilding initiatives typically seek to advance:

- **Trust and Cooperation.** Collaborative environmental projects foster trust and cooperation between adversaries, laying the foundation for peace and stability.
- **Conflict Prevention and Resolution.** By managing shared natural resources and addressing environmental issues collaboratively, communities and nations can reduce tensions and prevent conflicts arising from competition over natural resources, environmental degradation, and resource scarcity.
- **Environmental Justice.** Environmental justice ensures fair and equitable treatment of all individuals, including future generations, by considering their needs and rights in the development, implementation, and enforcement of environmental policies and practices.
- **Inclusive Governance.** Environmental peacebuilding involves inclusive decision-making that engages diverse stakeholders, including marginalized groups, ensuring their voices are heard and needs addressed.

Fig. 17.2 Core goals of environmental peacebuilding initiatives. (*Note*. Figure 17.2 lists core goals guiding environmental peacebuilding initiatives. These goals are not necessarily mutually inclusive and can independently contribute to the success of an initiative)

- **Resilience and Adaptation.** Strengthening environmental resilience and promoting adaptive strategies help communities cope with environmental challenges, reducing future conflict likelihood.
- **Sustainable Development.** Environmental peacebuilding promotes sustainable development by managing natural resources to benefit both the environment and local communities. Figure 17.2 lists the core goals that environmental peacebuilding initiatives typically seek to advance.

17.4 Theoretical Foundations of Environmental Science and International Relations

To explore how environmental science and international relations contribute to environmental peacebuilding, we must examine their theoretical foundations (Bennett and Chapin 2018; Sommer and Fassbender 2024). This understanding of key concepts, principles, and perspectives enables us to identify effective strategies for fostering environmental peace and stability. By delving into their theoretical underpinnings, we can uncover synergies and complementarities to address complex environmental challenges and promote sustainable, long-lasting peace.

17.4.1 Key Principles of Environmental Science

Environmental science is grounded in the study of natural systems and the interactions between organisms and their environment (Bennett and Chapin 2018; Odum 1971). It rests on principles from ecology, sustainability, and systems thinking to address complex environmental challenges. Drawing on Withgott and Laposata (2012), some key principles informing environmental science are:

- **Climate Science:** The study of climate systems, including the causes and effects of climate change, and the development of strategies to mitigate and adapt to its impacts (National Aeronautics and Space Administration n.d.).
- **Ecology:** The study of interactions between organisms and their environment, focusing on ecosystems, biodiversity, and the flow of energy and nutrients (Begon et al. 2006).
- **Environmental Ethics:** The philosophical study of moral values and principles concerning the environment, including the rights of nonhuman entities and human responsibilities towards nature (Callicott and Frodeman 2009).
- **Sustainability:** Meeting present needs without compromising future generations, framed around economic, environmental, and social sustainability (United Nations n.d.).
- **Systems Thinking:** Viewing problems as part of an overall system rather than isolated incidents. It emphasizes the interconnectedness of components and considers how changes in one part can affect the whole (Meadows 2008).

17.4.2 Key Perspectives in International Relations

International relations examines the dynamics of global politics, especially the interplay between nations, non-state actors, and international institutions (Kydd 2015; Mingst and Arreguín-Toft 2016). Informed by diverse theoretical perspectives like constructivism, liberalism, and realism, the field analyzes and addresses issues of cooperation, conflict, and power globally. Drawing on Reus-Smit and Snidal (2008), some key perspectives informing international relations are:

- **Constructivism:** The role of ideas, norms, and identities in shaping international relations, arguing that state behavior is influenced by social constructs and shared beliefs (Adler 2019).
- **Feminism:** Examines how gender shapes international relations, advocating for the inclusion of women's experiences and voices in global politics (True 2018).
- **Liberalism:** Emphasizes cooperation, democracy, and international institutions as means to achieve peace and stability, advocating for economic interdependence and multilateralism (Ikenberry 2018).

- **Marxism:** Analyzes international relations through the lens of economic inequality and class struggle, highlighting the impact of capitalism on global politics (Rupert and Smith 2016).
- **Realism:** Views international relations as a struggle for power among self-interested states, emphasizing national security and military strength (Mearsheimer 2018).

These theoretical foundations provide a framework for understanding the complex interactions between environmental issues and international relations, highlighting the importance of cross-disciplinary approaches to address global challenges.

17.5 Disciplinary Insights from Environmental Science and International Relations

17.5.1 Environmental Science: Key Concepts and Contributions

Environmental scientists have long studied how human activities impact ecosystems and how environmental degradation can lead to social unrest. From deforestation in the Amazon to water shortages in the Middle East, the link between environmental stress and conflict is well-documented (Bennett and Chapin 2018). Environmental science provides insights into peacebuilding's ecological aspects, such as conservation strategies, natural resource management, and the role of biodiversity in promoting stability and resilience. By revealing how environmental changes can trigger or escalate conflicts, it offers strategies to mitigate these risks through sustainable management and restoration efforts (Sommer and Fassbender 2024).

Environmental science uniquely contributes to the research question by providing crucial insights into the natural systems and processes underpinning environmental peacebuilding (Sommer and Fassbender 2024). Drawing on Cunningham and Cunningham (2022), here are key ways in which environmental science can inform environmental peacebuilding initiatives:

- **Climate Science.** Insights from climate science identify the impacts of climate change on conflict regions. Understanding challenges like increased droughts or rising sea levels allows scientists to recommend adaptive measures to mitigate impacts and reduce conflict risk.
- **Ecological Understanding.** Environmental science offers a deep understanding of ecosystems, biodiversity, and organism-environment interactions. This knowledge is essential for developing strategies to manage resources sustainably and restore conflict-degraded ecosystems.

- **Environmental Ethics.** Environmental science often incorporates environmental ethics, guiding decision-making regarding nonhuman entities' rights and humans' responsibilities towards nature. This ethical framework supports inclusive and equitable environmental peacebuilding initiatives.
- **Environmental Monitoring.** Advanced monitoring techniques allow scientists to track environmental changes in real-time. This data assesses ecosystem health, detects early signs of stress, and implements timely interventions to prevent conflict.
- **Restoration Ecology.** Environmental science includes restoration ecology, focusing on rehabilitating degraded environments. Restoring ecosystems can rebuild livelihoods, promote social stability, and provide a foundation for long-term peace in post-conflict regions.
- **Sustainable Resource management.** Environmental science provides methodologies for sustainably managing resources like water, forests, and minerals. Effective resource management prevents resource scarcity, often a conflict driver, and promotes equitable access, fostering community cooperation.

17.5.2 International Relations: Key Concepts and Contributions

International relations has traditionally focused on state interactions but increasingly recognizes the importance of non-state actors and transnational issues like the environment (Kydd 2015). Researchers examine how environmental issues influence conflict, international cooperation, and state behavior (Barnett 2001). The concept of environmental security highlights how environmental challenges can undermine national security and stability. International relations provides a framework for understanding the role of diplomacy, international organizations, and treaties in addressing environmental conflicts and promoting peace.

International relations uniquely contributes to answering the research question by providing insights into the economic, political, and social dynamics of environmental peacebuilding (Sommer and Fassbender 2024). Drawing on Bodansky et al. (2007), here are key ways in which international relations can inform environmental peacebuilding initiatives:

- **Conflict Prevention and Resolution.** International relations provides frameworks for conflict prevention and resolution, focusing on diplomacy, mediation, and negotiation. This helps design strategies to de-escalate tensions and promote peaceful coexistence.
- **Diplomatic Negotiations.** International relations helps understand how diplomatic negotiations and treaties manage shared environmental resources and address conflicts. For example, international agreements on water rights can help prevent disputes over water scarcity.

- **Economic Interdependence.** International relations appreciates the role of economic interdependence in promoting peace and stability. By fostering trade and economic cooperation, international relations help create conditions conducive to environmental peacebuilding.
- **Global Governance.** International relations explores the structures and mechanisms of global governance, including international environmental regimes and conventions. These mechanisms establish norms, standards, and regulations guiding environmental management and peacebuilding efforts.
- **Human Security.** International relations emphasizes the concept of human security, which includes environmental security as a key component. This perspective underscores the importance of protecting individuals and communities from environmental threats and ensuring their well-being.
- **International Cooperation.** International relations highlights the importance of international cooperation in addressing global environmental challenges. This includes the role of international organizations, such as the United Nations, in facilitating collaboration and coordination to mitigate environmental conflicts.
- **Power Dynamics.** International relations examines power dynamics and the influence of various state and non-state actors on environmental policies. Understanding these dynamics is crucial for developing effective environmental peacebuilding initiatives that account for different stakeholders' interests and actions.

17.5.3 Comparative Analysis of Environmental Science and International Relations

Naturally, given their distinct disciplinary origins, environmental science and international relations diverge significantly in their investigative approaches (Barnett 2018).

17.5.3.1 The Practice of Environmental Science

Environmental science is driven by cross-disciplinary collaboration, empirical and data-driven approaches, a focus on natural systems, and quantitative analysis to address environmental challenges (Cunningham and Cunningham 2022):

- **Cross-Disciplinary Collaboration.** Environmental science integrates knowledge from biology, chemistry, physics, and geology to understand complex environmental phenomena. This approach is essential for addressing multifaceted environmental issues.
- **Empirical and Data-Driven.** Environmental science relies on empirical data and scientific methods to study natural systems. Researchers conduct field stud-

ies, experiments, and use advanced technologies like remote sensing and geographic information systems to collect and analyze data.

- **Focus on Natural Systems.** Environmental science aims to understand interactions within natural systems, including ecosystems, climate systems, and the impact of human activities. The goal is to develop solutions for conservation, sustainability, and restoration.
- **Quantitative Analysis.** Environmental science employs quantitative analysis to measure environmental variables, model ecological processes, and predict future scenarios. This includes statistical analysis, mathematical modeling, and simulation.

17.5.3.2 The Practice of International Relations

International relations is driven by case studies and comparative analysis, focusing on governance and diplomacy, understanding political and social contexts, and using qualitative and theoretical analysis (Dunne et al. 2021; Mingst and Arreguín-Toft 2016).

- **Case Studies and Comparative Analysis.** International relations frequently uses case studies and comparative analysis to examine instances of international cooperation, conflict, and diplomacy. This approach helps identify patterns, draw lessons, and develop theories.
- **Focus on Governance and Diplomacy.** International relations investigates international governance mechanisms, the role of diplomacy, and the impact of international policies and agreements. The goal is to develop strategies for conflict prevention, resolution, cooperation, and promoting global stability.
- **Political and Social Context.** International relations emphasizes political and social contexts in shaping international interactions. Researchers examine the influence of cultural factors, ideologies, and power dynamics on global politics.
- **Qualitative and Theoretical Analysis.** International relations often relies on qualitative analysis and theoretical frameworks to study the behavior of states, international organizations, and non-state actors. Researchers analyze historical events, refer to policy documents, and conduct interviews to understand motivations and strategies in the international arena.

17.5.3.3 Contrasting Investigative Methods: Environmental Science and International Relations

Environmental science and international relations approach the research question from different angles, reflecting their unique focal points, guiding principles, integrative approaches, methodologies, and perspectives (Barnett 2018). Key areas of contrast include:

- **Focus Areas.** Environmental science converges on natural systems, ecological processes, and the physical environment. It aims to understand the natural world and develop solutions for environmental conservation, sustainability, and restoration. Conversely, international relations concentrates on the behavior of states, international organizations, and non-state actors in the global political arena. It seeks to understand the dynamics of power, cooperation, and conflict among these actors.
- **Guiding Principles.** Environmental science is grounded in scientific principles and emphasizes objectivity, empirical evidence, and the interconnectedness of natural systems. The goal is to develop scientifically-based solutions for environmental challenges. International relations is rooted in political theories and emphasizes the role of power dynamics, diplomacy, and governance. The goal is to develop strategies for conflict prevention, resolution, international cooperation, and global stability.
- **Integrative Approaches.** Environmental science integrates knowledge from various scientific disciplines such as biology, chemistry, and geology. International relations draws from political science, sociology, economics, and history to understand complex global interactions.
- **Research Methods.** Environmental science uses empirical data, field studies, laboratory experiments, and quantitative analysis to study environmental phenomena. Techniques like remote sensing, geographic information systems, and ecological modeling are commonly employed. International relations relies on theoretical frameworks, historical analysis, case studies, and qualitative methods. Researchers analyze policy documents, conduct interviews, and use comparative analysis to understand international interactions.
- **Viewpoints.** Environmental science views environmental challenges through the lens of ecological impact and sustainability, emphasizing preserving biodiversity and maintaining healthy ecosystems. International relations examines environmental challenges through the lens of political and economic impact, considering how environmental issues affect state behavior, international relations, and global security.

Given their differing focus areas, guiding principles, integrative approaches, research methods, and viewpoints, it is not surprising that the solutions proposed by environmental science and international relations vary significantly. In the context of environmental peacebuilding, environmental science typically offers technical and scientific solutions such as conservation strategies, ecosystem restoration, and sustainable resource management, prioritizing the environmental root causes of conflict (Millennium Ecosystem Assessment 2005). In contrast, international relations usually suggest policy recommendations, diplomatic strategies, and frameworks for international cooperation, focusing on managing political and social dynamics to prevent and resolve conflicts (Conca and Dabelko 2002). By understanding these divergent approaches, we can better appreciate each discipline's unique contributions to environmental peacebuilding and, where possible, combine them to create more comprehensive and effective initiatives.

17.6 Practical Implications of Integrating Disciplinary Insights

The integration of insights from environmental science and international relations in environmental peacebuilding initiatives has several practical implications (Conca and Dabelko 2002). Key implications include:

- **Effective Community Engagement**
  - **Capacity Building.** Insights from both disciplines support capacity-building initiatives that strengthen local communities' and institutions' ability to manage environmental resources sustainably. This reduces dependency on external assistance and promotes long-term resilience.
  - **Inclusive Participation.** Environmental science highlights involving local communities in environmental management, ensuring their knowledge and needs are considered. International relations emphasize stakeholder engagement and participatory approaches in decision-making. Together, these disciplines promote inclusive participation, empowering communities, and enhancing the legitimacy of environmental peacebuilding initiatives.
- **Enhanced Understanding of Conflict Drivers.** Combining the ecological knowledge of environmental science with the political and social insights of international relations helps develop a comprehensive understanding of environmental conflicts' underlying drivers. This holistic analysis can identify the root causes and dynamics of conflicts, leading to more targeted and effective interventions.
- **Improved Policy and Strategy Development**
  - **Adaptive Management.** Environmental science contributes methodologies for monitoring and assessing environmental conditions, while international relations provide frameworks for diplomatic engagement and conflict prevention or resolution. Synergizing the two disciplines enables adaptive management approaches based on real-time data and changing political contexts.
  - **Informed Decision-Making.** Integrating environmental science and international relations allows for creating policies and strategies informed by scientific data and political realities. This ensures that peacebuilding initiatives are grounded in empirical evidence and are politically feasible, increasing their likelihood of success.
- **Strengthened International Cooperation**
  - **Regional and Global Initiatives.** Synergizing environmental science and international relations enables developing regional and global initiatives addressing transboundary environmental issues. This fosters collaboration among neighboring countries and international organizations, reducing tensions and building trust.

  - **Shared Environmental Goals.** Environmental science emphasizes sustainable resource management, serving as common ground for cooperation among conflicting parties. International relations facilitate diplomatic endeavors and international agreements supporting collaborative environmental management, promoting peace and stability.

- **Sustainable Development and Resilience**
  - **Long-Term Solutions.** Environmental science provides technical solutions for ecosystem restoration and sustainable resource management, while international relations ensure these solutions are embedded within a supportive political framework. Harnessing both disciplines leads to sustainable development outcomes addressing environmental and social dimensions of conflict.
  - **Resilience Building.** By tackling the environmental root causes of conflict and promoting cooperative political strategies, integrated peacebuilding initiatives enhance communities' resilience to environmental and sociopolitical shocks. This reduces the risk of conflict re-emergence and contributes to lasting peace.

- **Technological Innovations**
  - **Advanced Monitoring and Evaluation.** Utilizing technologies such as data analytics, geographic information systems, and remote sensing enhances the monitoring and evaluation of environmental peacebuilding initiatives. These technologies provide accurate and timely data, enabling adaptive management and informed decision-making.
  - **Sustainable Technologies.** Research and implementation of sustainable technologies for climate adaptation, ecosystem restoration, and resource management significantly contribute to the success of environmental peacebuilding initiatives.

In summary, the integration of environmental science and international relations in environmental peacebuilding initiatives offers a robust and multifaceted approach to addressing the complex challenges posed by environmental conflicts. Figure 17.3 synergizes key concepts and contributions from environmental science and international relations, aiming to improve outcomes for environmental peacebuilding initiatives.

17.7 Challenges of Cross-Disciplinary Approaches to Environmental Peacebuilding

Cross-disciplinary approaches face significant real-world challenges (Matthew et al. 2002). These challenges often arise from the distinct theoretical foundations and disciplinary perspectives inherent to each field. Effectively addressing these

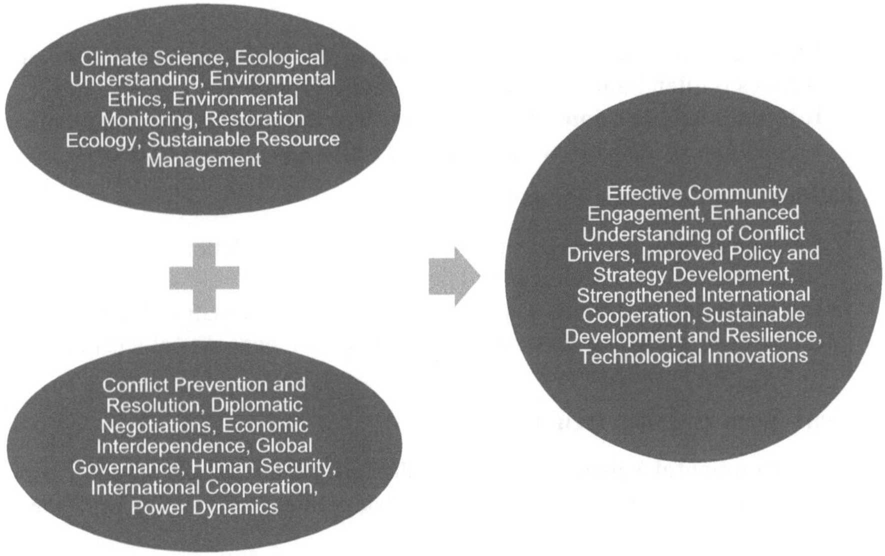

Fig. 17.3 Integrating disciplinary insights: synergizing environmental science and international relations

issues necessitates targeted capacity building, policy reforms at the disciplinary level, or customized actions tailored to specific situations. There are meaningful opportunities for future research. Among the key challenges are:

- **Complexity of Environmental Issues**
  - Environmental challenges like biodiversity loss, climate change, and pollution are complex and interconnected, making them difficult to address holistically.
- **Cross-Disciplinary Collaboration**
  - Effective communication and collaboration between professionals from environmental science and international relations are crucial but challenging due to cultural resistance and institutional barriers as well as their differing communication platforms and styles, data collection methods, focal points, funding resources, guiding principles, integrative approaches, methodologies, perspectives, and power dynamics.
- **Cultural and Social Factors**
  - Environmental peacebuilding initiatives must consider the cultural and social context of the affected communities, but understanding and integrating local knowledge and practices can be challenging for external actors.

- **Data Availability and Quality**
  - **Environmental Science.** Accurate data on environmental conditions and changes is often limited, particularly in conflict zones.
  - **International Relations.** Political data, like information about diplomatic negotiations or state behaviors, can be sensitive and difficult to obtain.
- **Political and Economic Constraints**
  - **Environmental Science.** Implementing scientific solutions often requires significant financial resources and political will.
  - **International Relations.** Diplomatic efforts and international cooperation can be hampered by competing national interests, geopolitical tensions, and lack of trust between states.
- **Short-Term vs. Long-Term Goals**
  - **Environmental Science.** Scientific solutions often require long-term commitment and monitoring to be effective, but political and funding cycles may prioritize short-term results.
  - **International Relations.** Diplomatic strategies and agreements can be influenced by short-term political agendas, which may not align with the long-term goals of environmental peacebuilding.

17.8 End Notes

Environmental peacebuilding offers a holistic approach to conflict prevention, resolution, and sustainable development by leveraging the power of nature to foster cooperation and resilience. It paves the way for lasting peace and sustainability by addressing the root causes of conflict and meeting communities' needs for a healthy environment and sustainable livelihoods (Conca and Dabelko 2002; Matthew et al. 2002).

However, cross-disciplinary approaches are essential in environmental peacebuilding, drawing on diverse perspectives and expertise to tackle complex challenges more effectively. This integration fosters innovative, contextually relevant, and sustainable solutions, ensuring initiatives are robust and adaptable to various situations (Bennett and Chapin 2018; Ostrom 2009). By collaborating across disciplines, we can create comprehensive strategies that resolve conflicts, promote long-term peace, ensure environmental sustainability, and enhance community well-being.

This précis demonstrated that insights from environmental science and international relations are vital for understanding the complex interplay between environmental factors and conflict dynamics. By integrating these disciplinary perspectives, we can develop effective strategies for managing natural resources, promoting inclusive governance, and enhancing community resilience (Homer-Dixon 1999; Koubi et al. 2014).

References

Adler E (2019) World ordering: a social theory of cognitive evolution. Cambridge University Press
Barnett J (2001) The meaning of environmental security: ecological politics and policy in the new security era. Zed Books
Barnett J (2018) Environmental science and international relations: comparative approaches and methodologies. Cambridge University Press
Begon M, Townsend C, Harper J (2006) Ecology: from individuals to ecosystems, 4th edn. Blackwell Publishing
Bennett E, Chapin F (2018) Interactions between ecology and environmental science. Environ Sci Technol 52(14):7632–7641
Bodansky D, Brunnée J, Hey E (eds) (2007) The Oxford handbook of international environmental law. Oxford University Press
Callicott J, Frodeman R (2009) Encyclopedia of environmental ethics and philosophy, vol 1. Macmillan Reference USA
Conca K, Dabelko G (2002) Environmental peacemaking. Johns Hopkins University Press
Conca K, Wallace J (2009) Environment and peacebuilding in war-torn societies: lessons from the UN Environment Programme's experience with post-conflict assessment. Glob Gov 15(4):485–504
Cunningham W, Cunningham M (2022) Environmental science: a global concern, 15th edn. McGraw-Hill Education
Dresse C, Ide T, Ide R (2018) Environmental peacebuilding: a review of the empirical evidence and future research directions. J Peace Res 55(4):491–503
Dresse A, Fischhendler I, Nielsen J, Zikos D (2019) Environmental peacebuilding: towards a theoretical framework. Coop Confl 54(1):99–119
Dunne T, Kurki M, Smith S (eds) (2021) International relations theories: discipline and diversity, 5th edn. Oxford University Press
Goldstein J, Pevehouse J (2017) International relations, 11th edn. Pearson
Heffington E, Krampe C, Scheffran J (2019a) Environmental peacebuilding: a critical review of the literature. J Peace Res 56(5):679–695
Heffington S, Oler A, Tretler D (eds) (2019b) A national security strategy primer. National Defense University Press
Homer-Dixon T (1999) Environment, scarcity, and violence. Princeton University Press
Ide T (2021) Environmental peacebuilding: moving beyond resolving violence-ridden conflicts to sustaining peace. World Dev 139:105348
Ikenberry G (2018) A world safe for democracy: liberal internationalism and the crises of global order. Yale University Press
Koubi V, Spilker G, Böhmelt T, Bernauer T (2014) Do natural resources matter for interstate and intrastate armed conflict? J Peace Res 51(2):227–243
Krampe C, Swain A (2024) Environmental peacebuilding and conflict prevention: new insights and approaches. Int Environ Agreem 24(1):123–145
Krampe F, O'Driscoll D, Johnson M, Simangan D, Hegazi F, de Coning C (2024) Climate change and peacebuilding: sub-themes of an emerging research agenda. Int Aff 100(3):1111–1130
Kydd A (2015) International relations theory. Cambridge University Press
Matthew R, Halle M, Switzer J (2002) Conserving the peace: resources, livelihoods, and security. International Institute for Sustainable Development
Matthew R, Brown O, Jensen D (2009) From conflict to peacebuilding: the role of natural resources and the environment. United Nations Environment Programme
Meadows D (2008) Thinking in systems: a primer. Chelsea Green Publishing
Mearsheimer J (2018) The Great Delusion: liberal Dreams and International Realities. Yale University Press
Millennium Ecosystem Assessment (2005) Ecosystems and human well-being: synthesis. Island Press

Mingst K, Arreguín-Toft I (2016) Essentials of international relations, 7th edn. W. W. Norton & Company
National Aeronautics and Space Administration (n.d.) Climate change: how do we know? NASA. https://climate.nasa.gov/evidence/
Odum E (1971) Fundamentals of ecology, 3rd edn. W.B. Saunders Company
Ostrom E (2009) A general framework for analyzing sustainability of social-ecological systems. Science 325(5939):419–422
Peace Direct & Alliance for Peacebuilding (2020) Local peacebuilding: what works and why. Peace Direct
Reus-Smit C, Snidal D (eds) (2008) The Oxford handbook of international relations. Oxford University Press
Roskin M, Cord R (2019) Political science: an introduction, 14th edn. Pearson
Rupert M, Smith H (2016) Historical Materialism and Globalization. Routledge
Scoones I (1998) Sustainable rural livelihoods: a framework for analysis. IDS Working Paper 72.
Serrat O (2017) The sustainable livelihoods approach. In: Knowledge solutions: tools, methods, and approaches to drive organizational performance. Springer, pp 21–26
Serrat O (2023) Safeguarding Lake Chad. Unpublished manuscript, Georgetown University. https://www.researchgate.net/publication/376618564_Safeguarding_Lake_Chad
Serrat O (2024a) Environmental peacebuilding: nurturing hope with analytical persuasion. Unpublished manuscript, Georgetown University. https://www.researchgate.net/publication/382637520_Environmental_Peacebuilding_Nurturing_Hope_with_Analytical_Persuasion
Serrat O (2024b) Safeguarding the mother of waters: environmental peacebuilding in the Mekong River Basin. Unpublished manuscript, Georgetown University. https://www.researchgate.net/publication/382520591_Safeguarding_the_Mother_of_Waters_Environmental_Peacebuilding_in_the_Mekong_River_Basin
Slingerland E (2008) What science offers the humanities: integrating body and culture. Cambridge University Press
Sommer A, Fassbender W (2024) Cross-disciplinary approaches to environmental peacebuilding: bridging the gap between theory and practice. Environ Sci Policy 47(2):34–45
True J (2018) The Political Economy of Violence against Women. Oxford University Press
United Nations (n.d.) Sustainable development goals: what are the sustainable development goals? https://sdgs.un.org/goals
United Nations Environment Programme (2023) Environmental cooperation and peacebuilding: a guide for practitioners. United Nations Environment Programme
Withgott J, Laposata M (2012) Essential environment: the science behind the stories, 4th edn. Pearson

Chapter 18
Environmental Peacebuilding: Nurturing Hope with Analytical Persuasion

Abstract This précis highlights the emergence of environmental peacebuilding as a growing field of research and practice, referencing environmental degradation and despoilation, including climate change. Despite promising developments, there remains a critical need to understand the true impact of environmental peacebuilding efforts. The précis identifies this gap and recommends further research into the role of evaluation criteria. By addressing this, the précis aims to contribute to the development of more robust and impactful environmental peacebuilding strategies.

18.1 Transforming the Earth

Concern about environmental degradation and despoilation has grown since 1958, when scientists at the Mauna Loa Observatory in Hawaii began reporting atmospheric carbon dioxide levels. Over the following decades, numerous stories of environmental and climate change emerged. In 1962, Rachel Carson's groundbreaking book, *Silent Spring*, highlighted the dangers of pesticides to wildlife and human health, and is credited with launching the modern environmental movement (Carson, 1962). In 1970, millions of people in the United States participated in the first Earth Day on April 22, raising awareness about environmental issues and leading to the creation of the Environmental Protection Agency. In 1986, the catastrophic nuclear accident in Chernobyl, Ukraine brought global attention to the environmental and health risks associated with nuclear energy. In 1992, the Global Environmental Conference in Rio de Janeiro elevated sustainability as a guiding goal for the global economy. In 1997, the Kyoto Protocol committed its parties to reduce greenhouse gas emissions, marking a significant step in global efforts to combat climate change. In 2015, the Paris Agreement within the United Nations Framework Convention on Climate Change aimed to limit global warming to well below 2°C above pre-industrial levels. Today—due to factors such as consequence of population growth, industrialization, urbanization, agricultural practices, energy use, economic expansion, and overconsumption in self-styled advanced countries—biodiversity loss, deforestation, environmental migration, global warming, melting glaciers and ice

O. Serrat, *Anthropogenic Solutions for Climate Change*,
https://doi.org/10.1007/978-981-96-8654-4_18

caps, ocean acidification, overfishing, pollution, soil degradation, and water scarcity are widely discussed and frequently featured in the news.

Resource-intensive economies declared war on the environment. While history books often detail human conflicts, they typically overlook the environmental damage caused before, during, and after these conflicts (Serrat 2024a). Conflicts frequently erupt over resources such as diamonds, precious stones, land, agriculture, oil, minerals, shared water sources, and timber, causing further environmental harm (Global Environmental Justice Atlas 2024; Serrat 2024b). Scarcity drives conflict, but resource abundance can also spark it (Bhavnani 2009; Klare 2001). Environmental history has never made comfortable reading: Radkau (2008) argued it is ultimately the history of human hopes and fears. While our use, exploitation, and adaptation of the Earth have brought benefits (Frankopan 2023), the cumulative environmental abuses of the Anthropocene, especially due to human dependence on fossil fuels, appear increasingly beyond the ability of our governments and societies to manage, resulting in a tragedy of the commons (Hardin 1968; Maslin 2021; Ostrom 1990; Rich 2019). António Guterres, Secretary-General of the United Nations, has warned that states must transform natural resources from drivers of conflict into tools for development, fostering peace and cooperation, as environmental degradation and despoliation serve as "threat multipliers" to conflict (United Nations 2018). The European Union's Copernicus Climate Change Service (2024) recently reported that the global average surface air temperature on July 21, 2024 reached 17.09°C (62.76°F), the warmest since records began in 1940, and revealed that every month since June 2023 has ranked as the planet's hottest.

18.2 Environmental Pathways to Peace: A Growing Field of Research and Practice

"Solastalgia," a new concept, captures the emotional and psychological distress caused by environmental change (Albrecht 2005; Galway et al. 2019). A book on psychology for sustainability, first published eight years ago, was in its fifth edition three years ago (Scott et al. 2021). Emerging in the late 1990s, environmental peacebuilding is a rapidly growing, transdisciplinary field at the intersection of environment (including climate change), conflict, peace, and security: it aims to address both the environmental risks of conflict and the opportunities for peace (Conca 2022; Dalmer 2022; Dresse et al. 2019; Swain and Öjendal 2018). This focus is crucial as the environment has become a core issue in international politics, and the number of armed conflicts remains high. Environmental peacebuilding means to address the environmental consequences of war; the use of natural resources to finance armed conflict; the dynamics of disasters and conflict; environmental factors in peace negotiations; the potential for cooperation around shared natural resources (e.g., the Arctic Council, peace parks, transboundary water cooperation) (Serrat 2024c, d); and the links among post-conflict peacebuilding, climate resilience, and natural resource management. The wide scope of environmental peacebuilding is evidenced by the work of the Environmental Peacebuilding Association (https://www.environmentalpeacebuilding.org/), which offers a vast library of assessments, case studies, policy briefs, reports, and multimedia content online. In conclusion, environmental peacebuilding has the potential to foster both negative and positive peace: negative peace can be achieved by addressing conflict resources and implementing measures to cease conflicts (Jensen and Lonergan 2012); positive peace can be cultivated by creating an environment that promotes cooperation and integration, rendering the notion of conflict inconceivable (Bruch et al. n.d.).

18.3 From Hope …

Environmental peacebuilding offers hope by fostering and sustaining peace through environmental cooperation and management. Snyder's (1994) cognitive model articulated three components: goals (creating desirable and achievable objectives), pathways (figuring out ways to work through obstacles and reach those objectives), and agency (believing in one's ability to instigate change and achieve those objectives). Optimism fuels achievement, as nothing is possible without hope and confidence. For lasting impact, however, Ali (2007) suggested that measured hope should be combined with analytical persuasion.

18.4 … to Analytical Persuasion …

Considering the future of environmental peacebuilding, Brown and Nicolucci-Altman (2022) remarked that despite political acceptance of the connection between the environment and violent conflict, actions often fall short of the rhetoric. Additionally, viewing environmental issues as potential triggers for violent conflict can unintentionally frame the environment as a security threat. Some analyses of environmental peacebuilding have faced criticism for their conceptual and methodological shortcomings, possibly underestimating societal adaptability to change. Moreover, organizations involved in environmental peacebuilding lack geographic and sectoral diversity. And, marginalized groups such as women, indigenous peoples, and youth are frequently viewed as passive aid recipients rather than active change-makers and knowledge-holders. Ide et al. (2021) pointed to knowledge gaps that should be addressed in future research, including bottom-up approaches, gender, conflict-sensitive programming, the use of big data and frontier technology, and monitoring and evaluation. Other challenges in the field include climate change adaptation, coordination and collaboration, cultural sensitivity, funding and resources, legal and institutional barriers, and political will. Strategies for monitoring and evaluating the impact of environmental peacebuilding initiatives would undoubtedly help promote accountability and continuous improvement.

18.5 … with "Best Practice"

"Best practice" is a method, technique, or set of procedures preferred or considered standard because it delivers superior results. Typically established through experience and research, best practices are proposed for widespread adoption. Kerr (2006) emphasized the importance of leadership commitment, strategic alignment, process optimization, technology integration, performance measurement, knowledge management, and cultural change in driving continuous improvement. As environmental peacebuilding evolves into an emerging framework, much of its discourse has focused on the conditions for success and constraints, often across the three trajectories of technical, restorative, and sustainable environmental peacebuilding (Brown and Nicolucci-Altman 2022; Carius 2008; Dresse et al. 2019; Ide et al. 2021). Brown and Nicolucci-Altman (2022) have criticized existing analyses of environmental peacebuilding for being conceptually and methodologically inadequate.

Exploring the conditions for success and constraints in environmental peacebuilding, Carius (2008) highlighted the function of environmental cooperation as a mechanism for conflict transformation and peacebuilding, identified the political and social conditions necessary for environmental cooperation to contribute to peacebuilding, and examined the institutional requirements (e.g., capable institutions, forms of participation) that foster or restrict the peacebuilding impacts of transboundary environmental cooperation projects. However, Carius (2008) noted there is little knowledge about the concrete role and impacts of environmental peacebuilding because (a) cooperative transboundary environmental projects often require negotiations that avoid explicit mention of conflict or peacebuilding, leading to broad objectives; (b) lack of administrative capacity often leads nongovernmental stakeholders to take an active role, complicating evaluation; (c) oft-postulated "win–win" situations either do not exist or are difficult to achieve; (d) resource degradation and user conflicts are not the sole or primary causes of violent conflicts, making it difficult to analyze the extent to which environmental components contribute to conflict genesis; (e) environmental peacebuilding tends to be initiated when conflicts are not intense, making the impact of individual projects difficult to assess; and (f) monitoring and evaluation of peacebuilding impacts are typically not included in the planning or implementation of interventions.

18.6 Bridging the Gap: Applying Evaluation Criteria in Environmental Peacebuilding

The unstated conclusion in Carius (2008) was that assessing the concrete role and impacts of environmental peacebuilding is difficult and perhaps not worth attempting. Moreover, Carius (2008) focused on armed conflict without considering that environmental peacebuilding can also be a preventive measure involving proactive steps to manage shared resources and ecosystems to prevent disputes from escalating. In such instances, evaluating the impacts of anticipative environmental peacebuilding should not be impossible. Lastly, in both post-conflict and anticipative environmental peacebuilding, it is surely beneficial to design projects according to specific criteria rather than on the fly.

The Organisation for Economic Co-operation and Development (2019) established six criteria for evaluating projects, programs, and policies: (i) relevance (is the intervention doing the right thing?), (ii) coherence (how well does the intervention fit?), (iii) effectiveness (is the intervention achieving its objective?), (iv) efficiency (how well are resources being used?), (v) impact (what difference does the intervention make?), and (vi) sustainability (will the benefits last?). Why would these criteria not be applicable to environmental peacebuilding, possibly supplemented by a seventh: legitimacy (does the intervention have the appropriate and recognized authority or authenticity?). In the realm of environmental peacebuilding, it is increasingly clear that a comprehensive examination of the application of criteria for evaluation is long overdue. This lack of attention not only hinders our understanding of the true impact of environmental peacebuilding efforts but also limits our ability to improve and refine these initiatives (Serrat 2017).

References

Albrecht G (2005) Solastalgia: a new concept in human health and identity. PAN Partners

Ali S (2007) Introduction: a natural connection between ecology and peace? In: Ali S, Marton-LaFevre J (eds) Peace parks: conservation and conflict resolution. MIT Press, pp 1–16

Bhavnani R (2009) Scarcity, abundance, and conflict: a complex new world. The Whitehead J Dipl Int Relat 10(2):65–80

Brown O, Nicolucci-Altman G (2022) White paper on the future of environmental peacebuilding. Geneva Peacebuilding Platform, International Union for Conservation of Nature, PeaceNexus Foundation, Environmental Law Institute, Environmental Peacebuilding Association. https://reliefweb.int/report/world/white-paper-future-environmental-peacebuilding

Bruch C, Weinthal E, Johnson M, Ide T (n.d.) Towards a definition of environmental peacebuilding. Ecosystem for peace. https://www.ecosystemforpeace.org/compendium/towardsadefinitionofenvironmentalpeacebuilding

Carius A (2008) Environmental cooperation as an instrument of crisis prevention and peacebuilding: conditions for success and constraints. Paper presented at the 2006 Berlin Conference on the Human Dimensions of Global Environmental Change "Resource Policies: effectiveness, Efficiency, and Equity", Berlin, November 17–18.

Carson R (1962) Silent spring. Houghton Mifflin

Conca K (2022) Environmental peacebuilding: yesterday, today, and tomorrow. J Soc Encount 8(1):4–12
Copernicus Climate Change Service (2024) New record daily global average temperature reached in July 2024, July 21. https://climate.copernicus.eu/new-record-daily-global-average-temperature-reached-july-2024
Dalmer N (2022) Building environmental peace: the UN Environment Programme as a knowledge actor. Palgrave Macmillan
Dresse A, Fischhendler I, Nielsen J, Zikos D (2019) Environmental peacebuilding: towards a theoretical framework. Coop Confl 54(1):99–119
Frankopan P (2023) The earth transformed: an untold history. Knopf
Galway L, Beery T, Jones-Casey K, Tasala K (2019) Mapping the solastalgia literature: a scoping review study. Int J Environ Res Public Health 16(15):1–24
Global Atlas of Environmental Justice (2024) The EJAtlas. https://ejatlas.org/
Hardin G (1968) The tragedy of the commons. Science 162(3859):1243–1248
Ide T, Bruch C, Carius A, Conca K, Dabelko G, Matthew R, Weinthal E (2021) The past and future(s) of environmental peacebuilding. Int Aff 97(1):1–16
Jensen D, Lonergan S (eds) (2012) Assessing and restoring natural resources in post-conflict peacebuilding. Routledge
Kerr J (2006) The best practices enterprise: a guide to achieving sustainable world-class performance. J. Ross Publishing
Klare M (2001) Resource wars: the new landscape of global conflict. Henry Holt
Maslin M (2021) How to save our planet: the facts. Penguin Life
Organisation for Economic Co-operation and Development (2019) Evaluation criteria: adapted definitions and principles for use. https://one.oecd.org/document/DCD/DAC%282019%2958/FINAL/en/pdf
Ostrom E (1990) Governing the commons: the evolution of institutions for collective action. Cambridge University Press
Radkau J (2008) Nature and power: a global history of the environment. Cambridge University Press
Rich N (2019) Losing earth: a recent history. Farrar, Strauss and Giroux
Scott B, Amel E, Koger S, Manning C (2021) Psychology for sustainability, 5th edn. Routledge
Serrat O (2017) Learning from evaluation. In: Knowledge solutions: tools, methods, and approaches to drive organizational performance. Springer, pp 879–884
Serrat O (2024a) Operation earth: integrating eco-friendly practices into the military. Unpublished manuscript, Georgetown University. https://www.researchgate.net/publication/382250401_Operation_Earth_Integrating_Eco-Friendly_Practices_into_the_Military
Serrat O (2024b) Neo-Malthusian and cornucopian perspectives: must it be either/or? Unpublished manuscript, Georgetown University. https://www.researchgate.net/publication/381768461_Neo-Malthusian_and_Cornucopian_Perspectives_Must_It_Be_EitherOr
Serrat O (2024c) Preserving paradise: the heart of Borneo initiative. Unpublished manuscript, Georgetown University. https://www.researchgate.net/publication/381993897_Preserving_Paradise_The_Heart_of_Borneo_Initiative
Serrat O (2024d) Safeguarding the mother of waters: environmental peacebuilding in the Mekong River Basin. Unpublished manuscript, Georgetown University. https://www.researchgate.net/publication/382520591_Safeguarding_the_Mother_of_Waters_Environmental_Peacebuilding_in_the_Mekong_River_Basin
Snyder C (1994) The psychology of hope: you can get there from here. Free Press
Swain A, Öjendal J (eds) (2018) Routledge handbook of environmental conflict and peacebuilding. Routledge
United Nations (2018) States must transform natural resources from driver of conflict into development tool to foster peace, cooperation, Secretary-General tells Security Council, October 16. https://press.un.org/en/2018/sc13540.doc.htm

Chapter 19
Harnessing Evaluative Criteria for Environmental Peacebuilding

Abstract This précis presents a research concept for a study aimed at harnessing evaluative criteria for environmental peacebuilding. It outlines the reasons for undertaking the proposed study, states the problem to be examined, defines the research question, and articulates the conceptual framework with reference to the methods of data collection and analysis. Additionally, it lists the outcomes of a preliminary literature review, defines key terms, and specifies the expected outcomes.

19.1 Primary Reasons for Undertaking the Study

Imagine a world where peace is nurtured not through treaties or negotiations, but through our collective stewardship of Earth's natural resources. This is the essence of environmental peacebuilding, an emerging interdisciplinary field focused on integrating the management of environmental issues to support conflict prevention, mitigation, resolution, and recovery (Ide et al. 2021; Krampe and Swain 2021). Shared environmental challenges—such as climate change, deforestation, and water scarcity—can foster cooperation instead of conflict (Dresse et al. 2019; Swain Öjendal 2018). Addressing these challenges together, communities and nations can build trust, reduce tensions, and lay the foundation for lasting peace.

My interest in environmental peacebuilding developed while working on environmental management in Asia and the Pacific, a region home to 4.8 billion people in 2024 (nearly 60% of the total world population) (Worldometer 2024). In over a dozen countries, I observed the transformative effects of globalization from the 1990s—cultural exchange, economic growth, technological advancement—alongside the starker realities of social inequality. However, globalization has done much more: it has increased industrial activity, contributed to deforestation—which not only reduced biodiversity but also increased carbon emissions—encouraged resource depletion, led to waste generation, multiplied transportation emissions, spread invasive species, and accelerated climate change, leaving an indelible mark on our planet's climate and ecosystems (Serrat 2017a).

O. Serrat, *Anthropogenic Solutions for Climate Change*,
https://doi.org/10.1007/978-981-96-8654-4_19

The chronicles of environmental history offer a compelling case for environmental peacebuilding. Environmental history delves into the dynamic interplay between natural processes and human activities over time, scrutinizing how humans have exploited natural resources, the ensuing environmental consequences, and how people's priorities have shaped their interactions with the environment. Unfortunately, environmental history and conflict are often intertwined. Environmental changes can exacerbate tensions, and competition for natural resources can spark conflicts that, in turn, harm the environment (Cabot 2016; Richards 2003). Conflict, that is, a dispute or incompatibility, arises when the actual or perceived needs, interests, positions, and values of individuals or groups clash (Fisher 2022; Matthew et al. 2009). It can manifest overtly through open confrontations and disputes or simmer beneath the surface, causing tension.

The relationship between natural resources, the environment, and conflict is multidimensional and complex. However, three main pathways can be identified: natural resources can contribute to the outbreak of conflict, finance and sustain conflict, and undermine peacemaking (Matthew et al. 2009). Likewise, conflict also impacts the environment through three main pathways: there are direct impacts (from physical destruction of ecosystems and wildlife and the release of polluting and hazardous substances into the natural environment); indirect impacts (from coping strategies such as the liquidation of natural assets and overuse of marginal areas by local and displaced populations to survive socio-economic disruption and loss of basic services); and institutional impacts (from disruption of state institutions, initiatives, and mechanisms of policy coordination, leading to poor management, lack of investment, illegality, the collapse of positive environmental practices, and the diversion of financial resources away from investments in public infrastructure and essential services) (Matthew et al. 2009).

The environment has profoundly shaped human history, and our actions have left enduring impacts on the planet. Here are some examples of environment–human and human–environment interactions:

- **The Ancient Mediterranean (Various Periods, Mediterranean Region).** Extensive deforestation and soil erosion in the Mediterranean region severely undermined agricultural productivity and societal stability (Hughes 1994). Overgrazing by livestock and the expansion of agriculture led to significant landscape degradation, impacting the sustainability of ancient civilizations such as the Egyptians, Phoenicians, Greeks, and Romans.
- **The Mongols (Thirteenth–Fourteenth Centuries, Central Asia and Europe).** The movements and conquests of the Mongols were heavily influenced by climatic conditions (Krajick 2014). The Mongol Empire's expansion was facilitated by periods of favorable weather, which improved pastures for their horses. Conversely, harsh winters and droughts could drive migrations and invasions.
- **The Inca Empire (Fifteenth–Sixteenth Centuries, South America).** The Inca Empire's terraced fields were a marvel of agricultural engineering, preventing soil erosion, managing water resources, and creating arable land in mountainous terrain (D'Altroy 2014).

- **Mughal India (Sixteenth–Nineteenth Centuries, South Asia).** Resource management was crucial to the stability of the Mughal Empire (Richards 1993). The Mughals implemented sophisticated irrigation systems and land revenue policies to enhance agricultural productivity and sustain their large population.
- **The Industrial Revolution (Eighteenth–Nineteenth Centuries, Europe).** Driven by its insatiable need for raw materials and energy, the Industrial Revolution brought about significant environmental changes, including deforestation, air and water pollution, and urbanization (Stearns 2012). The shift to coal as a primary energy source led to widespread environmental degradation.
- **The U.S. Civil War (1861–1865, United States).** The movement of large armies during the U.S. Civil War led to deforestation, soil compaction, and the destruction of crops and infrastructure (Gates 1965). The war significantly disrupted agricultural production and altered land use patterns.
- **The Dust Bowl (1930s, United States).** Severe dust storms caused by poor agricultural practices and drought led to massive soil erosion and forced many families to abandon their farms (Worster 2004). The disaster underscored the urgent need for sustainable farming practices and soil conservation.
- **The Aral Sea Crisis (Twentieth Century, Central Asia).** An environmental disaster caused by the diversion of rivers for irrigation projects in the Soviet Union led to the dramatic shrinking of the Aral Sea (Micklin 2007). This resulted in the collapse of local fisheries, the release of toxic dust from the exposed seabed, and health problems for nearby populations due to soil and water contamination by pesticides and fertilizers.
- **The First World War (1914–1918, Global).** Extensive trench warfare during the First World War altered landscapes, causing deforestation and soil erosion (Tucker 2005)). The use of chemical weapons led to long-term soil and water contamination. Military activities caused farmland destruction and food shortages.
- **The Second World War (1939–1945, Global).** The Second World War affected diverse environments globally, including urban destruction from bombing campaigns (Laakkonen et al. 2017). Extensive logging and mining for war efforts led to deforestation and habitat loss. Atomic bombings caused radiation contamination with long-term effects. Advanced weaponry increased environmental destruction.
- **The Great Smog of London (1952, United Kingdom).** A severe air pollution event caused by industrial emissions and weather conditions, the Great Smog resulted in thousands of deaths and pointed to the urgent need for air quality regulations (Bell et al. 2004).
- **The Green Revolution (Mid-Twentieth Century, Global).** The Green Revolution involved the introduction of high-yield crop varieties and modern agricultural techniques, but also led to environmental issues such as soil degradation, water scarcity, and pesticide pollution (Evenson and Gollin 2003). The increased use of chemical fertilizers and irrigation had long-term ecological impacts.

- **Twentieth-Century Guerrilla Warfare (Various Periods, Global).** Twentieth-century guerrilla warfare often involved the strategic use of natural resources (Beckett 2001). Guerrilla fighters used forests, mountains, and other natural terrains for cover and sustenance, impacting local ecosystems and resource availability.
- **The Chernobyl Disaster (1986, Ukraine).** The catastrophic nuclear accident at the Chernobyl Nuclear Power Plant released large quantities of radioactive particles into the atmosphere, leading to severe environmental contamination (Marples 1988). The disaster caused long-term health issues for the local population, forced the evacuation of entire towns, and left a lasting impact on the surrounding ecosystems. The Chernobyl Exclusion Zone remains one of the most contaminated areas in the world, showing up the lasting consequences of nuclear accidents.
- **The Amazon Rainforest (Ongoing, South America).** The Amazon rainforest is a quintessential example of human–environment interaction (Fearnside 2005). Often referred to as the "lungs of the Earth," the Amazon plays a crucial role in absorbing carbon dioxide (CO_2) from the atmosphere. However, deforestation for agriculture, logging, and mining has significantly impacted this vital carbon sink, leading to increased greenhouse gas emissions and impairing global climate regulation. Deforestation in the Amazon has also had severe consequences for biodiversity, resulting in ecosystem disruption and species loss. Additionally, indigenous communities have faced profound cultural impacts, displacement, health issues, and violence and exploitation due to the destruction of their ancestral lands.

Among many other notable examples, the Mayan Civilization (ca. 2000 BCE–900 CE, Mesoamerica), the Black Death (1347–1351, Europe), the Little Ice Age (ca. 1300–1850, Northern Hemisphere), the Columbian Exchange (Fifteenth–Sixteenth Centuries, Global), the Great Famine (1315–1317, Europe), the Irish Potato Famine (1845–1852, Ireland), and the Three Gorges Dam (Twenty-First Century, China) illustrate the complex and dynamic interactions between human societies and the environment throughout history. Tainter (1988) and Ponting (1991), two influential figures in the field of environmental history, were among the early scholars to examine the reasons behind the collapse of complex societies. Their work drew attention to the diminishing returns on investments in complexity and the overuse of natural resources.

While the impact of humans on nature and nature's influence on human societies are nothing new under the sun, the negative linkages between human and environmental issues became increasingly pronounced in the twentieth century due to escalating energy use, surging greenhouse gas emissions, and ballooning population growth (McNeill 2000; McNeill and Engelke 2014). Francopan (2023) observed that "[t]he last decades of the [20th] century saw massive and relentless growth in demand for natural resources and raw material" (p. 607) and suggested that we had reached the edge of ecological limits around 1990, 30–40 years ago. In 2000, the term "Anthropocene" was proposed to describe the current era in Earth's history, where human activity has become the dominant force shaping the planet in what McNeill and Engelke (2014) labeled a massive uncontrolled experiment (Crutzen and Stoermer 2000). This era is characterized by the impacts of human activity on biogeochemical cycles, biodiversity loss—which Kolbert (2014) compared to a Sixth Extinction—climate change, deforestation, land use changes, ocean acidification, pollution, resource depletion, soil degradation, and urban heat islands. Radkau (2008) believed that understanding the evolving relationship between humanity and nature is essential for comprehending world history. In 2024, who could disagree with Radkau (2008) that environmental policy has become a global concern?

Conversely, environmental history has the potential to highlight the positive outcomes that arise from shared environmental challenges in, say, biodiversity conservation, climate change adaptation and mitigation, community development, disaster risk reduction, energy and infrastructure, natural resource management, agriculture and food security, urban planning and development, or water resource management. For instance, transboundary water management agreements have often laid the groundwork for broader environmental peacebuilding efforts, such as the Indus Waters Treaty (1960, India and Pakistan), the Rhine Action Program (1987, Europe), and the Nile Basin Initiative (1999, Africa). At the global level, the Kyoto Protocol (1997) and the Paris Agreement (2015) are notable examples of existential undertakings designed to reduce greenhouse gas emissions and combat climate change. Appendix 1 provides curated lists of large-scale and smaller-scale environmental peacebuilding initiatives.

Environmental history can contribute to a more hopeful and proactive approach to addressing the pressing environmental issues of our time. "As we face a precarious future, learning lessons from the past has never been more important," Francopan (2023, back cover) reasoned. By learning from the past, we can develop strategies that not only mitigate environmental degradation but also promote peace and cooperation in a globalized world. Appendix 2 probes the interconnections between environmental history and environmental peacebuilding, emphasizing that understanding root causes, learning from the past, addressing contemporary challenges, promoting sustainable practices, and fostering cooperation are crucial for creating peaceful societies.

According to Google Trends (https://trends.google.com/), global interest in environmental peacebuilding emerged as recently as July 2016. However, environmental peacebuilding as a distinct field of research began to take shape in the late 1980s (Dresse et al. 2019). Hardt and Scheffran (2019) noted that the field grew out of discussions on environmental conflict, resource wars and environmental security, and more recently, climate change as a trigger for risks and conflict. Lately, the focus has been on integrating environmental conflict and peacebuilding research, with an emphasis on local dimensions. Appendix 3 identifies the key foci and content of environmental conflict research over the years. The field continues to evolve, incorporating insights from disciplines such as ecology, international relations, and political science to address the complex interactions between environmental issues and peacebuilding efforts. Currently, environmental peacebuilding aims to develop strategies that leverage natural resources to promote cooperation, prevent or reduce tensions, build resilient and sustainable livelihoods, and establish sustainable peace.

Environmental peacebuilding aligns closely with several Sustainable Development Goals (SDGs), particularly SDG 13 (Climate Action), which emphasizes the need for urgent action to combat climate change and its impacts; SDG 16 (Peace, Justice, and Strong Institutions), which promotes peaceful and inclusive societies; and SDG 17 (Partnerships for the Goals), which underscores the importance of global partnerships in achieving sustainable development. These goals emphasize the importance of addressing climate change, fostering peaceful and inclusive societies, and enhancing global partnerships, all of which are integral to effective environmental peacebuilding (United Nations General Assembly 2015).

Pursuing research in environmental peacebuilding has never been more critical. Climate change is exacerbating existing vulnerabilities in conflict-affected societies, placing immense stress on livelihoods and severely impacting food, water, and energy security. Furthermore, climate change is likely to intensify competition over natural resources, potentially igniting conflicts in regions historically free from such tensions. As a threat multiplier, climate change disrupts ecosystems and resource availability, causing previously stable areas to face new pressures and disputes over arable land, water, and other natural resources. Mach et al. (2019) concluded that intensifying climate change will increase future risks of conflict. To address these escalating challenges, it is imperative to enhance the effectiveness of environmental peacebuilding initiatives. These initiatives must be robust, adaptive, and capable of fostering sustainable peace.

Several fundamental human values are at stake in the process of environmental peacebuilding, including community engagement, equity, human rights, peace and security, and sustainability. These values are essential for creating a just and stable society where both people and the environment can thrive.

- **Community Engagement.** Active participation and collaboration among community members are crucial for fostering a sense of ownership and responsibility towards environmental initiatives. This engagement ensures that diverse voices are heard and that solutions are tailored to the specific needs and contexts of communities.
- **Equity.** Ensuring fair access to natural resources is vital for addressing historical injustices and disparities. Equity in environmental peacebuilding means that all individuals, regardless of their background, have the chance to benefit from and contribute to sustainable development.
- **Human Rights.** Protecting and promoting human rights is fundamental to environmental peacebuilding. This includes the right to a healthy environment, access to clean air and water, and the protection of vulnerable populations from environmental harm.
- **Peace and Security.** Environmental degradation can exacerbate conflicts and threaten security. By addressing environmental issues, peacebuilding efforts can reduce tensions, prevent conflicts, and promote long-term stability.

- **Sustainability.** Sustainable practices ensure that natural resources are managed responsibly, preserving them for future generations. Sustainability in environmental peacebuilding involves balancing economic, environmental, and social considerations to achieve lasting positive outcomes.

By prioritizing community engagement, equity, human rights, peace and security, and sustainability through environmental peacebuilding, we can build societies where both people and the environment flourish, paving the way for a more harmonious and resilient future.

19.2 Statement of the Problem

Environmental peacebuilding is at the crossroads of urgency and innovation. As the global community grapples with the intertwined crises of environmental degradation, climate change, and conflict, the emergence of environmental peacebuilding offers a beacon of hope. However, this nascent field is characterized by limited evidence on effective practices and contextual applicability (Bruch et al. 2023a, b; Ide et al. 2021). This fundamental challenge underpins many of the field's other issues.

- **Environmental Projects with Peacebuilding Objectives.** Many environmental peacebuilding initiatives are primarily environmental projects with peacebuilding objectives, often conceptualized with insufficient consideration of their effectiveness in preventing conflict, resolving conflicts peacefully, or fostering peace.
- **Focus on Upward Accountability.** Monitoring of environmental peacebuilding initiatives focuses on upward accountability rather than organizational or field-wide learning.
- **Long-Term Impact Manifestation.** The impacts of environmental peacebuilding initiatives often manifest over long time horizons, making short-term capture of lessons learned challenging.
- **Measurement Difficulties.** The focus of environmental peacebuilding initiatives is on building resilient, adaptive systems, which presents measurement difficulties.
- **Multiplicity of Actors and Systems.** Environmental peacebuilding involves a diverse array of actors, including governments, nongovernmental organizations, local communities, international bodies, and private sector entities acting in ecological, economic, political, and social frameworks. While the inclusion of diverse stakeholders brings multiple perspectives and expertise, it also introduces complexity in coordination and communication.
- **Underdeveloped Theories of Change.** Theories of change for environmental peacebuilding initiatives are often underdeveloped, resulting in an evidence base that is largely anecdotal and deductive—derived logically from problem analysis but lacking empirical validation.

- **Urgent Timelines and Dynamic Situations.** Environmental peacebuilding initiatives frequently operate under urgent timelines and in dynamic, insecure situations, resulting in unintended impacts.

Reflecting on the conditions for success and constraints in environmental peacebuilding, Carius (2008) noted that there is limited knowledge about the concrete role and impacts of environmental peacebuilding due to several factors: (a) cooperative transboundary environmental projects often require negotiations that avoid explicit mention of conflict or peacebuilding, leading to broad objectives; (b) lack of administrative capacity often induces nongovernmental stakeholders to take an active role, complicating evaluation; (c) the oft-postulated "win–win" situations are difficult to achieve; (d) resource degradation and user conflicts are not the sole or primary causes of violent conflicts, making it difficult to analyze the extent to which environmental components contribute to conflict genesis; (e) environmental peacebuilding tends to be initiated when conflicts are not intense, making the impact of individual projects difficult to assess; and (f) monitoring and evaluation of peacebuilding impacts are typically not included in the planning or implementation of interventions.

Carius (2008) implicitly concluded that evaluating the specific role and impacts of environmental peacebuilding is challenging and may not be worthwhile (Serrat 2024a). However, Carius (2008) focused on armed conflict without considering that environmental peacebuilding can also be a preventive measure involving proactive steps to manage shared resources and ecosystems to prevent disputes from escalating. In such cases, assessing the impacts of anticipative environmental peacebuilding should be feasible. Finally, for both post-conflict and anticipative environmental peacebuilding, it is advantageous to design projects based on specific criteria rather than improvising.

A crucial limitation, also recognized by Carius (2008), likely explains the aforementioned issues: evaluative criteria are rarely used to assess the merit, value, or worth of environmental peacebuilding initiatives, creating a significant knowledge gap (McClain et al. 2022). In other words, despite the growing recognition of environmental peacebuilding as a crucial field, evaluative thinking remains conspicuously absent from its core practices (Serrat 2024a). This gap hinders the integration of lessons learned, the identification of areas for improvement, and the recognition of good practices (Serrat, 2017b, c). It complicates the comparison of outcomes across different initiatives and regions, making it harder to replicate or scale up successful strategies. It does not foster accountability and transparency, essential to ensuring that stakeholders can trust the processes and outcomes. Without a normative framework for their evaluation, the true potential of environmental peacebuilding initiatives will remain untapped, diminishing their overall effectiveness and impact.

Despite the hurdles, the time has come to infuse evaluative thinking into environmental peacebuilding design, implementation, and monitoring (Ernstorfer et al. 2016). Appendix 4 outlines the evaluation criteria and principles recommended by the Organisation for Economic Co-operation and Development (OECD) for assessing development interventions. The proposed research would aim to recommend sets of evaluative criteria, some of them newly formulated, that meet the specific needs and contexts of stakeholders, and can be used to design and assess environmental peacebuilding initiatives. These criteria would substantially enhance the effectiveness of such efforts by providing a robust framework for design, implementation, monitoring, and evaluation. By establishing clear standards, these criteria would facilitate the identification and promotion of good practices that can be replicated. By proposing these evaluative criteria, the proposed research would seek to improve the comparability and scalability of environmental peacebuilding initiatives, thereby enabling more coordinated and impactful approaches. The use of these criteria would lead to more consistent and effective initiatives, helping to mitigate conflicts over natural resources and promoting sustainable development and resilience in conflict-affected regions. Furthermore, by providing a clear framework for evaluation, this research would make a substantial contribution to the broader field of peace and conflict studies, supporting the development of more robust and impactful environmental peacebuilding strategies.

19.3 Research Question

The specific, measurable, achievable, relevant, and time-bound (SMART) research question that the proposed research would answer is: Considering their infrequent application in environmental peacebuilding initiatives, how might evaluative criteria be customized to address the distinct sociopolitical and ecological challenges of various contexts and improve outcomes?

19.4 Conceptual Framework

Unveiling the hidden social dynamics that shape our environmental practices, social constructivism (often described as interpretivism) provides a vital lens to understand the complexities of environmental peacebuilding. This theoretical perspective is particularly suitable for presenting a world view on how specific evaluative criteria might enhance the effectiveness and sustainability of environmental peacebuilding initiatives. Naturally, the ontology, epistemology, axiology, rhetoric, and methodology of social constructivism would underpin the research paradigm for the proposed research.

Adopting a social constructivist worldview in studying environmental peacebuilding enables a deeper exploration of how social realities and environmental practices are co-constructed through human interactions. Social constructivism emphasizes the importance of understanding the subjective meanings and shared beliefs that communities attach to their environment and peacebuilding efforts. This perspective clarifies that environmental peacebuilding is not just a technical or ecological endeavor but also a social process shaped by cultural, historical, and political contexts.

By focusing on how different groups perceive and engage with their environment, researchers can uncover diverse narratives and power dynamics influencing peacebuilding initiatives. For instance, in post-conflict regions, restoring natural resources often involves negotiating between various community groups with differing historical ties and cultural significances attached to the land. This approach provides a more holistic understanding of environmental peacebuilding, recognizing the role of human agency and the constructed nature of social and environmental realities. By acknowledging and addressing these diverse perspectives, environmental peacebuilding efforts can become more inclusive and effective.

The proposed research would employ qualitative methods to solve problems by examining everyday life, exploring lived experiences, adopting participants' perspectives, and identifying patterns in behaviors or phenomena. This approach not only solves problems but also contributes to theory building. Typically broad and exploratory, qualitative research questions may not include predefined variables, acknowledging that relationships can emerge throughout the research process without necessarily relying on hypotheses. Consequently, inductive reasoning (e.g., observation, pattern, tentative hypothesis formation, theory development) would be crucial. In the field of environmental peacebuilding, qualitative methods afford a depth of understanding that quantitative approaches often lack. By employing qualitative methods, researchers can delve into the nuanced and complex interactions between environmental factors and peacebuilding processes, uncovering insights that might otherwise remain hidden.

For the proposed research, case studies would emerge as the most effective qualitative approach. They enable an in-depth examination of specific instances of environmental peacebuilding, offering rich, contextual insights that can reveal underlying mechanisms and dynamics. Through detailed narratives and comprehensive analyses, case studies can uncover the unique ways in which environmental initiatives contribute to peace across various settings. This approach not only enhances our theoretical understanding but also informs practical strategies for policymakers and practitioners working in the field of environmental peacebuilding.

Using case studies in qualitative research to investigate the use of evaluative criteria in environmental peacebuilding initiatives offers several compelling advantages:

- **Contextual Understanding.** Environmental peacebuilding initiatives are highly context-specific. Case studies enable researchers to explore the unique characteristics and dynamics of each case, providing insights that might be overlooked by other research methods.
- **Flexibility.** Case studies offer flexibility in research design and data collection methods, allowing researchers to adapt their approach based on the specific needs and constraints of their studies. This is particularly useful in the dynamic and often unpredictable field of environmental peacebuilding.

- **In-Depth Understanding.** Case studies allow for a comprehensive and detailed examination of complex phenomena within their real-world context. This is particularly valuable in environmental peacebuilding, where initiatives often involve intricate interactions between environmental, political, and social factors.
- **Longitudinal Insights.** Case studies can effectively capture longitudinal data, providing insights into the evolution and impact of peacebuilding initiatives over time. This is crucial for understanding the long-term effects and sustainability of these initiatives.
- **Multiple Data Sources.** Case studies typically involve the collection and analysis of data from various sources, such as interviews, observations, documents, and artifacts. This triangulation of data enhances the reliability and validity of the findings, offering a more holistic view of the evaluative criteria used in peacebuilding initiatives.
- **Practical Implications.** Case studies can provide practical insights and recommendations for policymakers and practitioners involved in environmental peacebuilding. By understanding what works and what does not in specific contexts, stakeholders can make more informed decisions and improve the effectiveness of their initiatives.
- **Theory Development.** By examining specific cases in detail, researchers can develop and refine theories related to evaluative criteria in environmental peacebuilding. This can lead to the identification of good practices and the development of more effective evaluation frameworks.

In summary, adopting a social constructivist worldview in environmental peacebuilding allows for a deeper understanding of how social realities and environmental practices are co-constructed through human interactions. This perspective emphasizes the importance of subjective meanings, shared beliefs, and the role of cultural, historical, and political contexts, best investigated by case studies of environmental peacebuilding initiatives.

Considering the following limitations, the proposed research could reasonably review between three and five case studies, that is one in each continent:

- **Number of Environmental Peacebuilding Initiatives.** There are numerous environmental peacebuilding initiatives worldwide, each with unique contexts and outcomes. Reviewing a few representative case studies would allow for a comprehensive understanding without overwhelming the scope of the proposed research study.
- **Vast Geographical Distribution.** Environmental peacebuilding initiatives are spread across different regions, each with distinct environmental and sociopolitical challenges. Selecting case studies from diverse geographical areas would ensure a well-rounded analysis and establish the global relevance of the proposed research.
- **Difficulty of Assembling Data.** Gathering detailed and reliable data for case studies can be challenging due to factors such as limited access to information and language barriers. Focusing on a manageable number of case studies allows for thorough data collection and analysis.

- **Time Constraints.** Limiting the number of case studies would ensure that each one can be examined in depth within the available timeframe, leading to a more robust and insightful thesis.

By balancing these factors, the proposed research can provide a meaningful contribution to the field of environmental peacebuilding while maintaining academic rigor and feasibility. The proposed research's regional coverage on environmental peacebuilding might include:

- **Latin America.** Countries such as Brazil and Colombia have rich biodiversity and face deforestation and land use conflicts. Studying initiatives in this region can showcase the role of environmental conservation in peacebuilding.
- **Middle East.** This region is known for its geopolitical tensions and water scarcity issues. Case studies from countries such as Egypt, Israel, Jordan could illustrate how environmental cooperation can contribute to peacebuilding.
- **South Asia.** Countries like Bangladesh and India face significant environmental challenges, including water scarcity and natural disasters. Examining peacebuilding efforts in this region can lay bare the intersection of environmental and social issues.
- **Southeast Asia.** Countries such as Indonesia and the Philippines deal with issues such as deforestation, natural disasters, and resource management (Serrat 1994). Case studies from this region can provide examples of community-based environmental peacebuilding efforts (Serrat 2024b).
- **Sub-Saharan Africa.** This region is the subject of numerous environmental peacebuilding initiatives, particularly in areas affected by resource conflicts and climate change. A case study from, say, Chad, Kenya, or Rwanda could provide valuable insights.

Appendix 5 presents a preliminary version of the conceptual framework for the proposed research. Appendix 6 details the rationale for using the case study method under the proposed research, contrasting it with other research designs.

19.5 Preliminary Literature Review

A literature review helps determine whether a topic is worth studying and provides insights into defining the scope of related areas of inquiry (Creswell 2014). To assess the viability of the topic, I conducted a search on Google Scholar for peer-reviewed articles published in English, without restricting the publication date. The search term "environmental peacebuilding" yielded only 136 results, with 93 (approximately 68%) published since 2020. Additionally, I searched Google Scholar for "environmental peacebuilding" and "evaluation" using the Boolean operator "AND." This combined search yielded only 68 results, with 43 (about 63%) published since 2020. Most of these articles referred to evaluation superficially. These limited results confirm a knowledge gap, establish the relevance and significance of the proposed research, and suggest that it would contribute to a small but growing corpus.

One of the most cogent references to evaluation in relation to environmental peacebuilding can be found in Ide et al. (2021). This article provided a comprehensive review of the field of environmental peacebuilding and singled out monitoring and evaluation as a key area for future research. It identified knowledge gaps that should be addressed in future research, including bottom-up approaches, gender, conflict-sensitive programming, the use of big data and frontier technology and emphasized the need for robust evaluation methods to assess the effectiveness of environmental peacebuilding initiatives.

More recently, the Environmental Law Institute in Washington, D.C. released a primer and a toolkit on monitoring and evaluation of environmental peacebuilding (Bruch et al., 2023a, b). Referring to the notable challenge of linking the environmental and peacebuilding dimensions, the toolkit focuses on monitoring and only recommends consideration of the OECD's (2023) evaluation criteria as a guide, preferring to underscore the importance of theories of change and advocating for the collection of data to build a picture of cause-and-effect relationships.

Past the scant peer-reviewed articles mentioned above, the literature review would scrutinize primers and toolkits such as Bruch et al. (2023a, b) for insights on the use of evaluation criteria for environmental peacebuilding. At a glance, however, other interested parties such as the OECD and the CDA Collaborative have focused on evaluation of conflict prevention and peacebuilding activities, not evaluation of environmental peacebuilding *per se*, with the former recommending pell-mell relevance, effectiveness, efficiency, impact, sustainability, linkages, coverage, and consistency with conflict prevention and peacebuilding values, and the second opting to concentrate exclusively on the evaluation criteria of effectiveness (CDA Collaborative Learning Projects 2016). The literature review would investigate and elucidate further. This would help refine the research question to ensure it addresses the identified gap and is feasible to answer via a detailed research plan.

19.6 Definitions of Key Terms

Constitutive definitions or interpretations of specific terms that are central to the proposed research and would be used throughout—expressly those that have multiple definitions in the literature, might be considered vague, or have a contextually related meaning—are listed below. These key terms are essential for understanding the concepts, methodologies, and findings that would be presented in the proposed research.

- *Climate Adaptation.* The process of adjusting to the ongoing and expected impacts of climate change. It involves implementing strategies to reduce vulnerability and enhance resilience to climate-related risks (International Panel on Climate Change 2014).

- *Climate Security*. The concept that climate change and related environmental stressors can exacerbate existing vulnerabilities and tensions, potentially leading to conflict. It emphasizes the need to address climate change as a security issue (Scheffran et al. 2012).
- *Conflict Resources*. Natural resources that are extracted, exploited, and traded to finance armed conflict, often leading to environmental degradation, despoilation, and prolonged violence. Examples include minerals, oil, and timber (Le Billon 2001).
- *Disaster Diplomacy*. The use of disaster response and recovery efforts to promote cooperation and reduce tensions between conflicting parties. It demonstrates how joint efforts in managing natural disasters can lead to improved relations and peacebuilding (Kelman 2012).
- *Ecological Diplomacy*. The use of diplomatic efforts to address ecological issues and promote environmental cooperation. It involves negotiations and agreements to manage shared ecosystems and resolve ecological conflicts (Conca 2006).
- *Ecosystem Services*. The conditions and processes by which natural ecosystems and their species sustain and support human life. They include "provisioning services" (e.g., fiber, food, timber, water); "regulating services" that affect, say, climate, disease, floods, wastes, and water quality; "cultural services" that provide aesthetic, recreational, and spiritual benefits; and "supporting services" such as nutrient cycling, photosynthesis, and soil formation (Matthew et al. 2009).
- *Environmental Diplomacy*. The use of diplomatic efforts to address environmental issues and promote cooperation between countries or groups. It involves negotiations, treaties, and agreements to manage shared natural resources and resolve environmental conflicts (Susskind 1994).
- *Environmental Peacebuilding*. "A meta-framework comprising multiple approaches and pathways by which management of environmental issues is integrated in and can support conflict prevention, mitigation, resolution, and recovery" (Bruch et al. 2023a, p. 8; Ide et al. 2021).
- *Environmental Security*. The concept that environmental degradation and resource scarcity can pose significant threats to national and global security. It underscores the need to address environmental issues to prevent conflicts and ensure stability (Homer-Dixon 1999).
- *Evaluation*. "The systematic and objective assessment of a planned, ongoing, or completed intervention, its design, implementation, and results" (OECD 2023, p. 31). The aim is to describe relevance, coherence, effectiveness, efficiency, impact, and sustainability (OECD 2024).
- *Evaluative Thinking*. The ongoing process of questioning, reflecting, learning, and modifying (Ernstorfer et al. 2016).
- *Livelihood*. "The capabilities, assets, and activities required for a means of living. A livelihood is sustainable when it can cope with and recover from stresses and shocks and maintain or enhance its capabilities and assets both now and in the future, while not undermining the natural resource base" (Chambers and Conway 1992, p. 9).

- *Natural Resources*. Materials or substances that occur in nature and can be used for economic gain, such as fertile land, hydrocarbons, metals, minerals, stones, timber, water, and wildlife. Renewable resources are naturally replenished at a rate that matches their consumption. Nonrenewable resources exist in a fixed amount or cannot be regenerated at a rate comparable to their consumption (Matthew et al. 2009).
- *Natural Resource Management*. The sustainable utilization and conservation of natural resources such as land, water, and biodiversity. Effective natural resource management is crucial for preventing resource-related conflicts and promoting environmental peacebuilding (Ostrom 1990).
- *Peacebuilding*. "[T]he identification and support of measures needed for transformation toward more sustainable, peaceful relationships and structures of governance, in order to avoid a relapse into conflict. The four dimensions of peacebuilding are: socio-economic development, good governance, reform of justice and security institutions, and the culture of justice, truth and reconciliation" (Matthew et al. 2009, p. 7).
- *Peacekeeping*. A political and military activity involving a presence in the field, with the consent of the parties involved, to implement or monitor arrangements for controlling conflicts (e.g., cease-fires, separation of forces) and resolving them (e.g., partial or comprehensive settlements). It also includes protecting the delivery of humanitarian aid (Matthew et al. 2009).
- *Peace Ecology*. An interdisciplinary approach that examines the interactions between ecological systems and peace processes. It focuses on how environmental sustainability and ecological health contribute to long-term peace and stability (Swain 2015).
- *Post-Conflict Reconstruction*. The process of rebuilding and rehabilitating a society after a conflict, including restoring infrastructure, governance, and social services. Integrating environmental considerations into post-conflict reconstruction is essential for sustainable peace (Barakat 2005).
- *Resource Governance*. The frameworks and processes through which natural resources are managed and regulated. Effective resource governance is crucial for preventing conflicts and promoting sustainable development (Young 2002).
- *Sustainable Development*. "Development […] that meets the needs of the present without compromising the ability of future generations to meet their own needs" (World Commission on Environment and Development 1987, para. 27). It encompasses economic, social, and environmental dimensions.
- *Transboundary Cooperation*. Collaboration between neighboring countries or regions to manage shared natural resources and address environmental challenges. It is a key aspect of environmental peacebuilding, as it fosters cooperation and reduces the risk of conflict over natural resources (Wolf 2007).
- *Water Diplomacy*. The use of diplomatic efforts to manage and resolve conflicts over shared water resources. It emphasizes the importance of cooperation and negotiation in ensuring equitable and sustainable water use (Islam and Susskind 2012).

19.7 Expected Outcomes

The outcomes of the proposed research would be expected to include:

- **Development of Tailored Evaluative Criteria.** The proposed research would create specific evaluative criteria that consider the unique sociopolitical and ecological challenges of environmental peacebuilding initiatives in different contexts. These evaluative criteria would be designed to be flexible and adaptable.
- **Improved Monitoring and Evaluation Frameworks.** The proposed research would provide insights into the development of robust monitoring and evaluation frameworks that can accurately assess the impact of environmental peacebuilding initiatives. These frameworks would serve to advance performance measurement, accountability, learning and improvement, decision-making, impact assessment, resource allocation, policy formulation and advocacy, and stakeholder engagement.
- **Increased Awareness and Advocacy.** The proposed research would raise awareness about the importance of tailored evaluative criteria in environmental peacebuilding and advocate for their adoption. This would be attempted primarily by disseminating research findings through publications and presentations.
- **Policy Recommendations.** The proposed research would formulate policy recommendations that support the implementation of tailored evaluative criteria in environmental peacebuilding initiatives. The aim of these recommendations would be to guide policymakers, practitioners, and stakeholders in designing and implementing more effective and sustainable interventions.

By delivering these outcomes, the proposed research would contribute to a deeper understanding of how to effectively customize evaluative criteria to address the unique challenges of environmental peacebuilding, ultimately leading to more successful and sustainable initiatives.

Appendix 1: Curated Lists of Large-Scale and Smaller-Scale Environmental Peacebuilding Initiatives

Environmental peacebuilding initiatives have emerged as essential efforts to address the intertwined challenges of environmental degradation and conflict. The following curated list showcases significant large-scale environmental peacebuilding initiatives, showcasing diverse approaches in leveraging environmental stewardship to achieve peace and stability. The list does not evaluate the efficacy of the initiatives included.

Africa

- **Great Green Wall Initiative.** Launched in 2007, this African-led initiative aims to combat desertification and land degradation across the Sahel region by creating a mosaic of green and productive landscapes.
- **Lake Victoria Basin Commission.** Established in 2003, this intergovernmental organization promotes sustainable management and development of the Lake Victoria Basin, involving Kenya, Uganda, and Tanzania, to address environmental and socio-economic challenges and ensure regional stability.
- **Nile Basin Initiative.** Established in 1999, this partnership among Nile Basin countries promotes cooperative management and development of the Nile River's water resources to ensure peace and stability in the region.
- **Peace Parks Foundation.** Established in 1997, this initiative aims to create transboundary conservation areas, known as peace parks, to promote biodiversity conservation and foster peace and cooperation among African nations.
- **Transfrontier Conservation Areas.** Formalized in 1999, these cross-border protected areas span multiple countries, promoting biodiversity conservation and fostering cooperation among nations. One notable example is the expansive Kavango–Zambezi Transfrontier Conservation Area, which stretches across Angola, Botswana, Namibia, Zambia, and Zimbabwe.

Asia

- **Aral Sea Basin Program.** Initiated in the 1990s, this program involves Kazakhstan, Kyrgyzstan, Tajikistan, Turkmenistan, and Uzbekistan working together to address the environmental and socio-economic challenges of the Aral Sea Basin.
- **Central Asia Regional Economic Cooperation Program.** Launched in 1997, this program includes 11 countries (Afghanistan, Azerbaijan, People's Republic of China, Georgia, Kazakhstan, Kyrgyz Republic, Mongolia, Pakistan, Tajikistan, Turkmenistan, Uzbekistan) working together to address common environmental and economic issues, focusing on sustainable development and regional cooperation.
- **Hindu Kush–Himalayan Monitoring and Assessment Program.** Launched in 1983, this initiative aims to monitor and assess the environmental health and social well-being of the Hindu Kush–Himalayan region, promoting cooperation among the countries sharing this critical ecosystem.
- **Indus Waters Treaty.** Signed in 1960 between India and Pakistan, this treaty facilitates the cooperative management of the Indus River system, helping to prevent water-related conflicts between the two countries.
- **Mekong River Commission.** Formed in 1995, this intergovernmental organization works to manage and develop the Mekong River Basin in a sustainable and

cooperative manner, involving Cambodia, Laos, Thailand, and Vietnam (Serrat 2024c).

Europe

- **Alpine Convention.** Established in 1991, this international treaty involves eight Alpine countries (Austria, France, Germany, Italy, Liechtenstein, Monaco, Slovenia, Switzerland) working together to protect the natural environment and promote sustainable development in the Alpine region, fostering regional collaboration.
- **Baltic Sea Action Plan.** Launched in 2007, this initiative involves countries around the Baltic Sea working together to restore the ecological balance of the sea and promote regional cooperation and peace.
- **Danube River Protection Convention.** Signed in 1994, this agreement among Danube River Basin countries aims to ensure the sustainable and equitable use of water resources and to prevent conflicts over water management.
- **European Green Belt Initiative.** Launched in 2003, this initiative seeks to create a network of protected areas along the former Iron Curtain, promoting biodiversity conservation and fostering peace and cooperation among European countries.
- **Mediterranean Action Plan.** Initiated in 1975, this action plan involves 21 countries bordering the Mediterranean Sea, focusing on environmental protection and sustainable development to foster peace and cooperation in the region.

North America

- **Great Lakes Water Quality Agreement.** Initially signed in 1972 between the United States and Canada, this agreement aims to restore and protect the water quality of the Great Lakes, fostering cooperation and preventing conflicts over shared water resources.
- **International Joint Commission.** Formed in 1909, this intergovernmental commission addresses and resolves issues concerning the shared waters between the United States and Canada, promoting cooperation and preventing conflicts over water resources.
- **Tri-National Commission for Environmental Cooperation.** Established in 1994, this commission involves Canada, Mexico, and the United States collaborating on environmental conservation and sustainable development to enhance regional cooperation.
- **US–Mexico Border Environment Cooperation Commission.** Established in 1993, this commission addresses environmental issues along the US–Mexico

border, promoting sustainable development and cooperation between the two countries.

- **Yukon River Inter-Tribal Watershed Council.** Founded in 1997, this council brings together indigenous tribes and First Nations from Alaska and Canada to protect and preserve the Yukon River watershed through cooperative management and sustainable practices.

Oceania

- **Coral Triangle Initiative.** Launched in 2009, this initiative involves six countries (Indonesia, Malaysia, Papua New Guinea, the Philippines, Solomon Islands, and Timor-Leste) collaborating to protect marine biodiversity and promote sustainable development in the Coral Triangle region.
- **Great Barrier Reef Marine Park Authority:** Established in 1975, this authority manages the Great Barrier Reef Marine Park, promoting sustainable use and conservation of the reef while fostering cooperation among stakeholders.
- **Micronesia Challenge.** Launched in 2006, this initiative involves five Micronesian countries (Guam, Marshall Islands, the Northern Mariana Islands, the Federated States of Micronesia, Palau) committing to effectively conserve at least 30% of nearshore marine resources and 20% of terrestrial resources by 2020.
- **Pacific Islands Climate Action Network.** Established in 2013, this network aims to address climate change impacts in the Pacific Islands, fostering regional cooperation and resilience-building among island nations.
- **Pacific Regional Environment Programme.** Established in 1982, this program promotes cooperation among 22 Pacific Island countries and territories to protect and manage the environment and natural resources.

South America

- **Amazon Cooperation Treaty Organization.** Founded in 1995, this intergovernmental organization promotes sustainable development and cooperation among Amazon Basin countries (Bolivia, Brazil, Colombia, Ecuador, Guyana, Peru, Suriname, Venezuela) to address environmental challenges and prevent conflicts over natural resources.
- **Andean Community of Nations.** Established in 1969, this initiative promotes regional integration and sustainable development among its four member countries (Bolivia, Colombia, Ecuador, Peru), addressing environmental and socio-economic challenges to foster peace and cooperation.
- **Gran Chaco Americano Initiative.** Launched in the 1990s, this initiative focuses on the sustainable management of the Gran Chaco region, involving

Argentina, Bolivia, Brazil, and Paraguay, to promote peace and cooperation through environmental conservation of the largest dry forest in the world and the second-largest forest biome in South America.
- **Pact for the Amazon.** Launched in 2019, this initiative involves Amazon Basin countries (Brazil, Bolivia, Colombia, Ecuador, Guyana, Peru, Suriname, Venezuela) working together to protect the Amazon rainforest, promote sustainable development, and prevent conflicts over natural resources.
- **Pantanal Regional Environmental Program.** Launched in 2008, this initiative focuses on the conservation and sustainable development of the Pantanal, the world's largest tropical wetland area, involving Brazil, Bolivia, and Paraguay.

There are many smaller-scale environmental peacebuilding initiatives that focus on local communities. The following curated list showcases five smaller-scale environmental peacebuilding initiatives from various continents, showcasing diverse approaches in leveraging environmental stewardship to achieve peace and stability. The list does not evaluate the efficacy of the initiatives included.

Africa

- **Upper Tana-Nairobi Water Fund (Kenya).** Initiated in 2015, this watershed management project works with local farmers to implement sustainable agricultural practices that protect the watershed and ensure a reliable water supply for Nairobi. It involves local communities in conservation efforts and conflict resolution related to water use.

Asia

- **Joint Environment Venture for South Asia (India, Pakistan).** Launched in 2022, this platform brings together environmental experts, think tanks, business leaders, and peacemakers from India and Pakistan. It aims to address shared environmental challenges such as melting glaciers, floods, and air pollution, fostering cross-border collaboration and innovative solutions for peace and environmental sustainability.

Europe

- **Community of Practice on Environment, Climate, Conflict, and Peace (Geneva, Switzerland).** Established in 2020 by the Geneva Peacebuilding Platform, this community of practice facilitates interaction on peacebuilding between different actors, institutions, and sectors. It aims to incubate collective initiatives and has grown into a community of more than 950 members. The community of practice focuses on addressing environmental causes of conflict and using collaboration around environmental issues to build peace.

North America

- **The Greening of Detroit (United States).** Launched in 1999, this initiative transforms vacant lots into community gardens and green spaces, fostering community engagement and improving urban environments. It aims to promote environmental sustainability and provide peaceful areas for community interaction.

Oceania

- **Transcend Oceania (Australia, Fiji, Solomon Islands).** Launched in 2022, this initiative focuses on the complex relationship between peacebuilding and climate change. It involves peacebuilders from the Pacific who participate in courses and workshops to explore how environmental stewardship can foster peace and stability. The initiative emphasizes developing strategic frameworks for peacebuilding, integrating arts and culture, and addressing the impacts of climate change on peace and security.

South America

- **Chocó Rainforest Conservation and Sustainable Livelihoods Initiative (Colombia).** Launched in 2010, this initiative focuses on conserving the Chocó Rainforest, one of the world's most biodiverse regions, while promoting sustainable livelihoods for local communities. It involves reforestation, sustainable agriculture, and eco-tourism projects that help protect the environment and provide economic opportunities for residents. By fostering cooperation and environmental stewardship, this initiative aims to mitigate conflicts over natural resources and promote peace and stability in the region.

Appendix 2: The Interconnections Between Environmental History and Environmental Peacebuilding

Introduction

The lessons of the past should guide us toward a future of peace and sustainability. This is the promise of intertwining environmental history with environmental peacebuilding. Environmental history and environmental peacebuilding offer insights into how human interactions with the environment shape peaceful societies. In their groundbreaking work, *Natural Enemy, Natural Ally*, Tucker and Russell (2004) compiled the first collection of essays examining the interplay between war and the environment. They contended that this facet of environmental history had been largely neglected due to the subdiscipline's roots in conservation history, which traditionally emphasized civilian over military matters (Bennett 2009; Tucker and Russell 2004). Building on this foundation, we can delve into historical contexts to uncover the underlying causes of environmental conflicts. By understanding these causes, we can devise strategies to foster peace through sustainable environmental practices. The following sections will explore how past experiences and lessons can inform and enhance current efforts in environmental peacebuilding.

Historical Context of Environmental Conflicts

Environmental history offers valuable insights into the patterns and triggers of environmental conflicts. For example, disputes over the Nile River between Egypt and its upstream neighbors, and conflicts over the Jordan River among Israel, Jordan, and Palestine, exemplify how water scarcity can escalate tensions (Pacific Institute 2024). Similarly, competition for high-value resources like diamonds, minerals, and oil has fueled civil wars in countries such as Angola, the Democratic Republic of Congo, and Liberia (Matthew et al. 2009). Additionally, climate change exacerbates resource scarcity, increasing the likelihood of conflicts over natural resources. For instance, contemporary droughts and desertification have intensified competition for water and land in regions like the Sahel, leading to conflicts between farmers and herders (Pacific Institute 2024). Environmental history also reveals how economic factors, weak state institutions, and lack of infrastructure can exacerbate resource-related tensions (United Nations 2024). By examining the historical context of environmental conflicts, we can identify patterns and triggers that highlight the role of natural resource management in supporting peace and stability.

Environmental Degradation and Social Unrest

Environmental history often reveals the impact of environmental degradation on social stability. Climate change, deforestation, and pollution have historically led to displacement, economic hardship, social unrest, and conflict. As early as the thirteenth century, for example, harsh climatic conditions, including extreme cold and drought, led the Mongol tribes on a series of military campaigns in search of better resources and more favorable living conditions (di Cosmo 2022). In the Amazon, deforestation has displaced indigenous communities and led to conflicts over land rights (Climate Transform 2022). The pollution of the Cuyahoga River in the United States, which caught fire multiple times, spurred significant environmental legislation and social activism (National Oceanographic and Atmospheric Administration 2024). Understanding the link between environmental degradation and social unrest is crucial for preventing conflicts and promoting stability.

The Role of Indigenous Knowledge

Environmental history can underscore the indispensable role of indigenous knowledge in fostering sustainable practices and building resilient, peaceful societies (Vizina 2019). These practices, often rooted in a deep understanding of local ecosystems and a holistic worldview, emphasize the interconnectedness of all living things. By incorporating traditional ecological knowledge into environmental peacebuilding efforts, we can enhance the effectiveness of these initiatives. This approach not only promotes respect for the environment but also fosters collaboration between different cultural groups, creating a more inclusive and culturally sensitive framework for addressing environmental challenges (Vizina 2019).

Building Resilience Through Environmental Restoration

Environmental history can illustrate the potential for environmental restoration to contribute to peacebuilding. Restoring degraded ecosystems creates opportunities for cooperation, rebuilds trust among communities, and provides economic benefits that support long-term peace (Sändig et al. 2024). Reforestation projects not only help in sequestering carbon and improving biodiversity but also bring together different community groups to work towards a common goal, fostering a sense of unity and collaboration (U.S. Department of Agriculture 2024). Similarly, the rehabilitation of wetlands can enhance water quality, provide habitat for wildlife, and support local economies through activities like fishing and ecotourism (Ramsar Convention on Wetlands 2021). These restoration efforts can serve as a foundation for building stronger, more resilient communities that are better equipped to handle future environmental challenges. By investing in environmental restoration, we can create a positive cycle where ecological health and social stability reinforce each other, ultimately contributing to lasting peace.

Economic Incentives for Environmental Peacebuilding

Historical analysis can reveal how economic incentives have influenced environmental management and conflict resolution. For example, the development of trade routes and resource extraction in colonial times often led to environmental degradation and conflicts. Understanding these historical economic drivers can help design modern incentives that promote sustainable practices and peace.

International Cooperation and Policy

Environmental history shows how international agreements and policies have shaped environmental outcomes. For instance, the history of the Montreal Protocol, which successfully addressed ozone depletion, provides lessons on how international cooperation can tackle global environmental issues. This historical perspective can inform current efforts to address climate change and environmental conflicts through international collaboration.

Technological Innovations

Historical advances in technology, such as the development of irrigation systems or early conservation efforts, have had significant impacts on environmental management. By studying these historical innovations, we can understand how technology has been used to mitigate environmental conflicts and apply these lessons to modern technological solutions.

Education and Awareness

Environmental history highlights the role of education and public awareness in shaping environmental policies and practices. For example, the environmental movement of the 1960s and 1970s, driven by increased public awareness, led to significant legislative changes. Understanding these historical movements can guide current efforts to educate and engage the public in environmental peacebuilding.

Gender and Environmental Peacebuilding

Historical analysis can reveal the gendered dimensions of environmental management and conflict. For instance, women have historically played crucial roles in managing natural resources and leading conservation efforts. By examining these historical roles, we can better understand how to empower women in contemporary environmental peacebuilding initiatives.

Lessons from Past Environmental Management

Historical case studies of successful and failed environmental management practices offer valuable lessons. For instance, the sustainable agricultural practices of the ancient Maya civilization, such as milpa farming, which involved crop rotation and forest management, can serve as a model for modern sustainable agriculture (Fisher 2024). Likewise, the use of decentralized irrigation systems by Zuni farmers in the Southwestern United States helped them survive long periods of drought (Fisher 2024). Ancient practices can be adapted to address current environmental challenges. For example, the revival of medieval Moorish irrigation technology has helped drought-stricken areas of Spain (Fisher 2024). On the other hand, the collapse of the Viking colonies in Greenland due to unsustainable farming practices underscores the dire consequences of poor environmental management (Fisher 2024). Through interdisciplinary approaches combining insights from environmental history, archaeology, and environmental science, the lessons learned from past successes and failures can inform the development of policies that promote sustainable resource management and conflict resolution.

Conclusion

By intertwining environmental history with environmental peacebuilding, we gain a profound understanding of the intricate relationship between humans and the environment. This understanding is essential for fostering sustainable and peaceful societies. Through the lessons of the past, we can identify and address the root causes of environmental conflicts, paving the way for a harmonious coexistence with nature. As we move forward, it is imperative that we integrate historical insights into our environmental policies and practices, ensuring that the mistakes of the past are not repeated. By doing so, we can build a future where peace and sustainability go hand in hand, creating a legacy of resilience and harmony for generations to come.

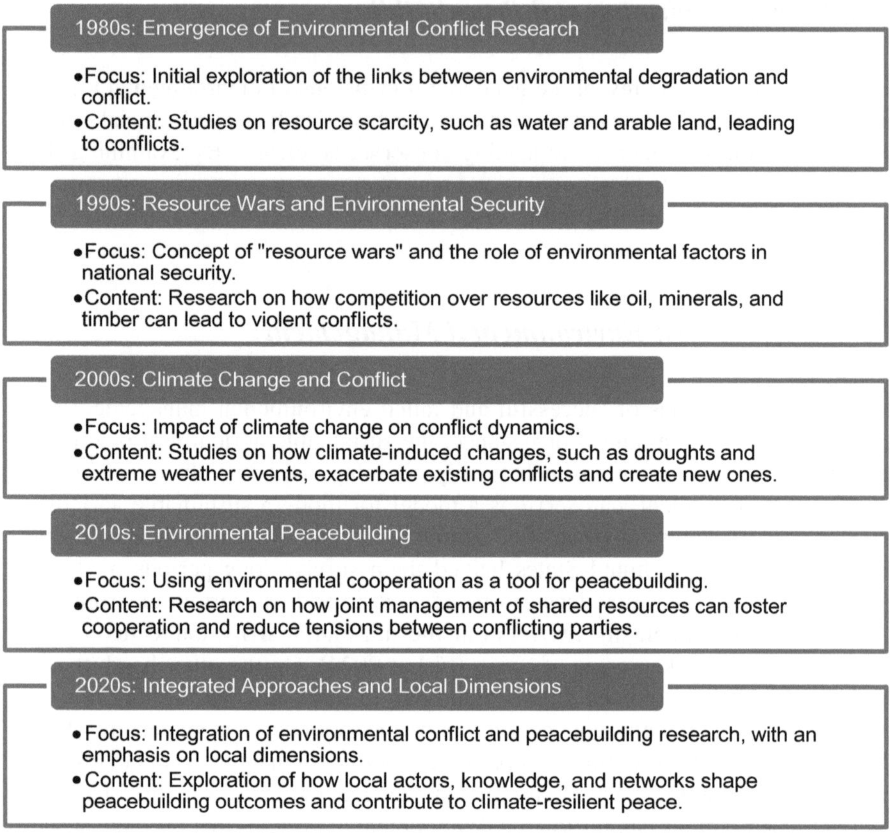

Fig. 19.1 Key foci and content of environmental conflict research over the years. (Source: Developed from Hardt and Scheffran (2019) and Sändig et al. (2024))

Appendix 3: Key Foci and Content of Environmental Conflict Research over the Years

Figure 19.1 summarizes the main themes and areas of focus in environmental conflict research, highlighting the evolution of the field over time.

Appendix 4: A Note on Evaluation Criteria

The Organisation for Economic Co-operation and Development (OECD) promotes the use of evaluation criteria to assess the merit, value, or worth of a development intervention (e.g., policy, strategy, program, project, activity). These criteria provide a structured framework for evaluative judgments and help ensure that evaluations are consistent, transparent, and useful.

Revisiting the OECD's Evaluation Criteria

The Development Assistance Committee (DAC) in the OECD first laid out the evaluation criteria in 1991. The DAC Network on Development Evaluation (EvalNet) revisited the definitions and use of the evaluation criteria in 2018–2019 and introduced adapted definitions for relevance, effectiveness, efficiency, impact, and sustainability, along with a new criterion, coherence (OECD, 2024). The changes aimed to:

- **Lay Out Guiding Principles.** New guiding principles were introduced to help evaluators apply the OECD's (2024) criteria more effectively. The update reflects the evolving landscape of international development and aims to enhance the quality and utility of evaluations.
- **Clarify Definitions.** The updated definitions of the OECD's (2024) are clearer and support more rigorous and nuanced analysis, including considerations of equity and synergies.
- **Introduce Coherence as a New Criterion.** The new criterion, coherence, was added to the OECD's (2024) evaluation criteria to evaluate how well the intervention fits with other interventions in the same context.
- **Contextual Adaptation.** The OECD's (2024) criteria are now intended to be used more thoughtfully and adjusted to the specific context of the intervention and the needs of the intended users of evaluations.

Principles for Using the OECD's (2024) Evaluation Criteria

- **Principle One.** The OECD's (2024) evaluation criteria should be applied with careful consideration to ensure high-quality and useful evaluations. They must be understood within the context of the specific evaluation, the intervention being assessed, and the stakeholders involved. The evaluation questions (what you are trying to find out) and what you intend to do with the answers should inform how the criteria are specifically interpreted and analyzed (OECD 2024).
- **Principle Two.** The application of the OECD's (2024) evaluation criteria should vary based on the evaluation's purpose. The criteria should not be applied mechanistically. Instead, they should be covered according to the needs of the relevant stakeholders and the context of the evaluation (OECD 2024).

Flexibility is essential. The Global Environment Facility (GEF), for example, uses the criteria of relevance, effectiveness, efficiency, and sustainability (but not impact) to assess projects because its evaluation policy of 2019 incorporated impact into the evaluation of effectiveness. In 2020, the GEF conducted its first-ever evaluation of GEF support in fragile and conflict-affected situations.

Six Lenses to Analyze an Intervention and Its Results

- **Relevance.** Is the intervention doing the right things? (OECD 2024). The criterion of relevance examines the extent to which an intervention's objectives and design respond to beneficiaries' needs, global priorities, and partners' policies (OECD 2024).
- **Coherence.** How well does the intervention fit? (OECD 2024) The criterion of coherence assesses the compatibility of an intervention with other interventions in the same context (OECD 2024).
- **Effectiveness.** Is the intervention achieving its objectives? (OECD 2024) The criterion of effectiveness measures the extent to which an intervention's objectives were achieved (OECD 2024).
- **Efficiency.** How well are resources being used? (OECD 2024) The criterion of assesses how economically resources/inputs (e.g., expertise, funds, time, etc.) were converted to results (OECD 2024).
- **Impact.** What difference does the intervention make? (OECD 2024) The criterion of impact looks at the broader effects of the intervention, including positive and negative, direct and indirect, and intended and unintended impacts (OECD 2024).
- **Sustainability.** Will the benefits last? (OECD 2024) The criterion of sustainability evaluates the likelihood of the benefits of an intervention continuing after external support has ended (OECD 2024).

The OECD's (2024) evaluation criteria are intended to play a normative role. Together, they describe the desired attributes of interventions: all interventions should be relevant to the context, coherent with other interventions, achieve their objectives, deliver results efficiently, and have lasting positive impacts (OECD 2019). The OECD's (2024) evaluation criteria are used to: (a) support accountability, including providing information to the public; and (b) support learning by generating and feeding back findings and lessons (OECD, 2019). Additionally, the OECD's evaluation criteria are to be used beyond evaluation for monitoring and results management, as well as for strategic planning and intervention design (OECD 2019).

Criteria to Assess and Evaluate Environmental Peacebuilding

Development interventions typically prioritize economic and social goals in agriculture, education, environmental and natural resource management, health, infrastructure, and public administration, usually in relatively stable environments. In these settings, the evaluation criteria of relevance, coherence, effectiveness, efficiency, impact, and sustainability are applicable. However, they should still be thoughtfully tailored to the needs of the relevant stakeholders and the context of the development intervention, as recognized by the OECD (2024). From a conflict prevention and peacebuilding perspective, not environmental peacebuilding *per se*, Van Brabant (2010) queried early the appropriateness of the OECD's evaluation criteria to peacebuilding work and recommended adaptation. Examining the criterion of relevance, for example, Van Brabant (2010) observed that an initiative that once was appropriate might not remain so as circumstances change in a fast-paced environment. Therefore, the criterion of relevance should also assess the responsiveness and flexibility of an initiative. To illustrate further, Van Brabant (2010) perceived that the criterion of sustainability elicits questions of commitment and implementation but also robustness in the face of challenges.

In contrast to development interventions, environmental peacebuilding initiatives often operate in contexts marked by conflict, instability, and the need for reconciliation. Although these contexts may not always be as extreme as those in conflict prevention and peacebuilding efforts, they still present significant challenges. Environmental peacebuilding initiatives aim to address environmental issues while fostering peace and cooperation among conflicting parties. The key differences between development interventions and environmental peacebuilding initiatives lie in their objectives, stakeholders, and sociopolitical landscapes. Therefore, environmental peacebuilding requires a nuanced approach that considers underlying tensions and the potential for environmental cooperation to build trust and reduce conflict. In contrast to development interventions, six drivers that frame environmental peacebuilding initiatives, over which peacebuilders have at best limited control, include:

- **Community Engagement.** Engaging local communities is crucial for the success of environmental peacebuilding initiatives. This involves actively involving community members in decision-making processes and ensuring their voices are heard and respected.
- **Conflict Intensity.** The severity and duration of a conflict significantly influence the approach to environmental peacebuilding. High-intensity conflicts may require more robust and immediate interventions, while lower-intensity conflicts might allow for more gradual and sustainable efforts.
- **Cultural Factors.** Cultural attitudes towards environmental conservation and peacebuilding play a vital role in shaping these initiatives. Understanding and respecting local customs, traditions, and beliefs is essential for gaining community support and ensuring the long-term success of the initiatives.

- **Economic Conditions.** Economic stability and available resources are essential for the effectiveness of environmental peacebuilding efforts. Adequate funding and economic opportunities can support the implementation and sustainability of these initiatives.
- **Environmental Degradation.** Addressing the extent of environmental damage and resource scarcity is a key focus of environmental peacebuilding. Efforts must be made to restore and protect natural resources to prevent further degradation and promote long-term environmental health.
- **Political Stability.** The level of political stability and governance in a region impacts the success of environmental peacebuilding initiatives. Stable political environments are more conducive to implementing and sustaining peacebuilding efforts, while unstable regions may face additional challenges. Table 19.1 illustrates how the OECD's evaluation criteria align with varying stakeholder needs and contextual factors, offering insights into their relative significance across different evaluative contexts.

Environmental peacebuilding summons contextual adaptation of evaluative criteria. Where the OECD's evaluation criteria are not fully appropriate, there may be a need to consider other evaluative criteria, such as those listed below:

- **Adaptability and Resilience.** The criterion of adaptability and resilience would evaluate the environmental peacebuilding initiative's ability to adapt to changing environmental and social conditions.
- **Conflict Sensitivity and Do No Harm.** The criterion of conflict sensitivity and do no harm would ensure that the environmental peacebuilding initiative does not exacerbate existing conflicts or create new ones.
- **Cultural Sensitivity.** The criterion of cultural sensitivity would ensure that the environmental peacebuilding initiative respects and integrates local cultural practices and knowledge.
- **Inclusivity and Participation.** The criterion of inclusivity and participation would assess the extent to which the environmental peacebuilding initiative involves and benefits all relevant stakeholders, including marginalized groups.

Table 19.1 Bearing of the OECD's evaluation criteria depending on stakeholder needs and contexts

	Community engagement	Conflict intensity	Cultural factors	Economic conditions	Environmental degradation	Political stability
Relevance	High	Medium	High	Medium	High	High
Coherence	High	Medium	High	Medium	High	High
Effectiveness	High	High	High	Medium	High	Medium
Efficiency	High	Medium	High	High	Medium	Medium
Impact	High	High	High	Medium	High	High
Sustainability	High	Medium	High	Medium	High	High

- **Legitimacy.** The criterion of legitimacy would assess the extent to which the environmental peacebuilding initiative is perceived to be appropriate, acceptable, and justifiable in the eyes of the stakeholders involved, as revealed by accountability, alignment with local norms and values, local ownership, stakeholder trust, and transparency.
- **Long-Term Environmental Impact.** The criterion of long-term environmental impact would assess, beyond medium-term outcomes, the long-term environmental benefits and sustainability of the environmental peacebuilding initiative.
- **Peace Writ Large.** The criterion of peace writ large would evaluate the broader impact on peace and stability in the concerned region, beyond the immediate goals of the environmental peacebuilding initiative.

By assessing the relevance of the OECD's evaluation criteria and combining them with one or several of the new criteria listed above, considering the specific needs and contexts of stakeholders, we can achieve a more comprehensive assessment of the merit, value, or worth of environmental peacebuilding initiatives, making it easier to replicate successful strategies. The proposed research would ascertain the relevance of these additional criteria, for instance by means of the literature review and the case studies. There would be a need for continued simplicity, by retaining a limited set of evaluative criteria and keeping the definitions coherent. Table 19.2 proposes evaluation criteria specifically designed to align with stakeholder needs and the contextual dynamics that define environmental peacebuilding initiatives.

Table 19.2 Using additional evaluation criteria depending on stakeholder needs and context

	Community engagement	Conflict intensity	Cultural factors	Economic conditions	Environmental degradation	Political stability
Adaptability & resilience	High	High	High	Medium	High	High
Conflict sensitivity & do no harm	High	High	High	Medium	High	High
Cultural sensitivity	High	Medium	High	Medium	High	High
Inclusivity & participation	High	Medium	High	Medium	High	High
Legitimacy	High	Medium	High	Medium	High	High
Long-term environmental impact	High	Medium	High	Medium	High	High
Peace writ large	High	High	High	Medium	High	High

Applying the Matrix

For each proposed environmental peacebuilding initiative, the selection of evaluative criteria would be assessed against the specific needs and contexts of stakeholders to determine their suitability and potential effectiveness. Additional considerations would be:

- **Adaptive Management.** Ensure the matrix allows for flexibility and adaptation based on changing conditions and feedback.
- **Continuous Learning.** Use the matrix as a tool for continuous learning and improvement, incorporating lessons learned from past initiatives.
- **Stakeholder Involvement.** Engage stakeholders in the evaluation process to ensure diverse perspectives and buy-in.

By systematically aligning evaluative criteria with prevailing conditions, one can better assess the potential success and impact of environmental peacebuilding initiatives.

A Note on Terminology

The terms "evaluation criteria" and "evaluative criteria" are often used interchangeably, but their connotations can vary slightly depending on the context:

- **Evaluation Criteria.** These are specific standards, such as the OECD's (2024), against which the merit, value, or worth of something (e.g., policy, strategy, program, project, activity) is measured. They define what is being measured and how it will be judged, the standards by which accomplishments of technical and operational effectiveness or suitability characteristics may be assessed, against which conformance, performance, and suitability of a technical capability, activity, product, or plan is measured.
- **Evaluative Criteria.** This term is more commonly used in academic or theoretical contexts and refers to the underlying principles or standards that guide the evaluation process. It involves the broader concepts or dimensions that are considered important in making judgments. For instance, in educational research, evaluative criteria might include validity, reliability, and fairness.

In summary, while both terms relate to the process of assessment, "evaluation criteria" are the specific, concrete standards used in an evaluation, whereas "evaluative criteria" refer to the broader, more abstract principles that inform the evaluation process.

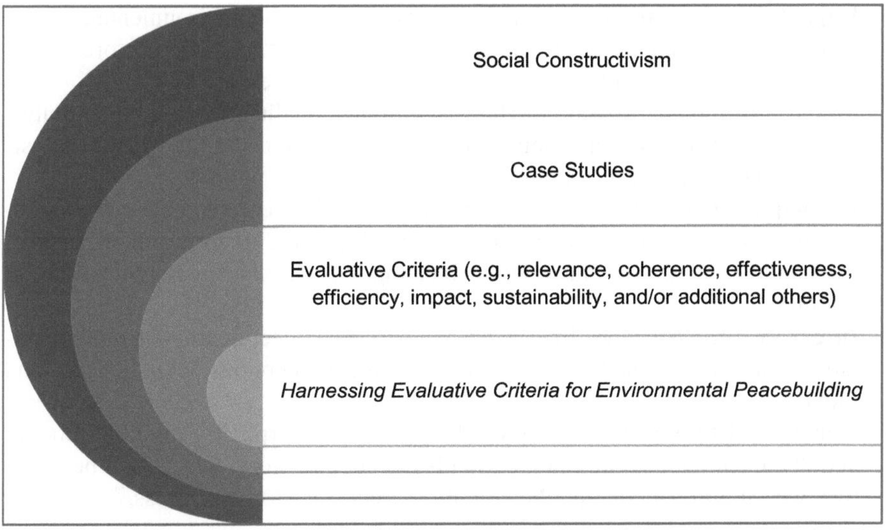

Fig. 19.2 Conceptual framework for the proposed research. (*Note.* Adopting a social constructivist perspective, the proposed research would review case studies of environmental peacebuilding initiatives to assess the integration of evaluative criteria in their design. Based on these case studies, the proposed research would tailor evaluative criteria to meet the specific needs and contexts of stakeholders)

Appendix 5: Conceptual Framework for the Proposed Research

Figure 19.2 visually outlines the key components and relationships within the proposed research, providing a clear and organized representation of its structure and methodology.

Appendix 6: Context Matters: Leveraging Case Studies in Environmental Peacebuilding Research

Potentially, several qualitative and quantitative methodologies other than case studies can be used to examine environmental peacebuilding initiatives. They include:

- **Comparative Analysis.** Comparative analysis juxtaposes different environmental peacebuilding initiatives to identify common factors of success or failure, helping to understand what works in different contexts.
- **Geospatial Analysis.** Geographic Information Systems and remote sensing analyze environmental changes and impacts over time, providing visual and spatial data to support assessments.

- **Impact Assessments.** Impact assessments appraise the environmental, social, and economic impacts of the initiatives, including both ex-ante (before implementation) and ex-post (after implementation) assessments.
- **Mixed Methods.** Mixed methods combine qualitative and quantitative approaches to provide a comprehensive assessment using statistical analysis alongside narrative descriptions.
- **Participatory Action Research.** Participatory action research engages community members and stakeholders in the research process to ensure their perspectives and knowledge are included, enhancing the relevance and applicability of findings.
- **Interviews and Surveys.** Interviews and surveys collect qualitative and quantitative data from stakeholders, helping gauge perceptions, experiences, and impacts.
- **Theory of Change.** The theory of change involves explicitly thinking through and documenting how an intervention is intended to work, why it will be effective, who it will benefit, and the conditions necessary for its success.

However, researchers looking to study environmental peacebuilding initiatives are likely to encounter several challenges:

- **Attribution of Outcomes.** It can be difficult to attribute observed changes directly to environmental peacebuilding initiatives, as many factors may influence the outcomes.
- **Complexity of Context.** Environmental peacebuilding initiatives often operate in post-conflict regions with complex environmental, political, and social dynamics.
- **Cultural Sensitivity.** Researchers must navigate cultural differences and ensure that their methods and approaches are culturally appropriate and respectful.
- **Data Availability and Quality.** Reliable data may be scarce, especially in conflict-affected areas.
- **Ethical Considerations.** Addressing ethical issues related to working in vulnerable communities is crucial.
- **Funding and Resources.** Limited funding and resources can constrain the ability to conduct thorough assessments.
- **Interdisciplinary Collaboration.** Environmental peacebuilding is inherently interdisciplinary: effective collaboration across various fields can be challenging.
- **Measuring Long-Term Impacts.** Environmental peacebuilding initiatives often aim for long-term outcomes, which can be difficult to measure within typical research study timeframe.
- **Security Concerns.** Conducting research in post-conflict regions can pose safety risks that may limit the scope and depth of assessments.
- **Stakeholder Engagement.** Ensuring meaningful participation from all relevant stakeholders, including marginalized groups, is crucial.

Given these challenges to the study of environmental peacebuilding initiatives, the qualitative and quantitative methods listed earlier betray certain limitations:

- **Lack of Contextual Depth.** Methods such as surveys and interviews may lack the depth and contextual understanding needed for complex and context-specific environmental peacebuilding initiatives.
- **Limited Flexibility.** Quantitative methods may not adapt well to the dynamic nature of environmental peacebuilding.
- **Inadequate Exploration of Complex Interactions.** Interviews and surveys may not capture the intricate interactions between social, environmental, and political factors.
- **Surface-Level Insights.** Quantitative methods often provide surface-level insights that may not fully capture the underlying mechanisms.
- **Challenges in Capturing Longitudinal Data.** Longitudinal studies can be challenging to implement using methods other than case studies.
- **Potential for Oversimplification.** Quantitative methods may oversimplify the complex realities by reducing them to numerical data.
- **Difficulty in Addressing Unique Cases.** Interviews and surveys may struggle to address unique, case-specific cases effectively.

In summary, while various research methods have their strengths, case studies are particularly well-suited for investigating the complex, context-specific nature of environmental peacebuilding initiatives. They offer the depth, flexibility, and comprehensive understanding needed to effectively explore and evaluate these initiatives. Their compelling advantages include contextual depth, flexibility, in-depth understanding, longitudinal insights, multiple data sources, practical implications, and potential for theory development.

References

Ali S (2007) Peace parks: conservation and conflict resolution. MIT Press

Barakat S (ed) (2005) After the conflict: reconstruction and development in the aftermath of war. I.B. Tauris

Beckett I (2001) Modern insurgencies and counter-insurgencies: Guerrillas and their opponents since 1750. Routledge

Bell M, Davis D, Fletcher T (2004) A retrospective assessment of mortality from the London smog episode of 1952: the role of influenza and pollution. Environ Health Perspect 112(1):6–8

Bennett J (2009) Natives and exotics: World War II and environment in the Southern Pacific. University of Hawaii Press

Bruch C, Woomer A, et al (2023a) Primer on monitoring and evaluation of environmental peacebuilding. Environmental Law Institute. https://m-and-e.environmentalpeacebuilding.org/assets/documents/11426452854f.pdf

Bruch C, Woomer A, et al (2023b) Toolkit on monitoring and evaluation of environmental peacebuilding. Environmental Law Institute. https://m-and-e.environmentalpeacebuilding.org/assets/documents/90a9ecd9f452.pdf

Cabot C (2016) Causal linkages between environmental change and conflict. In: Cabot C (ed) Climate change, security risks and conflict reduction in Africa. Springer, pp 123–145

Carius A (2008) Environmental cooperation as an instrument of crisis prevention and peacebuilding: conditions for success and constraints. Paper presented at the 2006 Berlin Conference on

the Human Dimensions of Global Environmental Change "Resource Policies: effectiveness, Efficiency, and Equity", Berlin, November 17–18.
CDA Collaborative Learning Projects (2016) Reflecting on peace practice (RPP) basics: a resource manual. CDA Collaborative Learning Projects. https://www.cdacollaborative.org/publication/reflecting-peace-practice-rpp-basics-resource-manual/
Chambers R, Conway G (1992) Sustainable rural livelihoods: practical concepts for the 21st century. IDS Discussion Paper 296. Institute of Development Studies
Climate Transform (2022) The history of deforestation: the past and origins, April 21. https://climatetransform.com/the-history-of-deforestation-the-past-and-origins/
Conca K (2006) Governing water: contentious transnational politics and global institution building. MIT Press
di Cosmo N (2022) Beyond boom and bust: climate in the history of medieval steppe empires. In: Idzebski A et al (eds) Perspectives on public policy in societal environmental crises: what the future needs from history. Springer, pp 177–190
Creswell J (2014) Chapter 2: review of the literature. In: Research design: qualitative, quantitative and mixed methods approaches. Sage Publications. Inc., pp 25–50
Crutzen PJ, Stoermer E (2000) The "Anthropocene". Global Change Newsletter 41:17–18
D'Altroy T (2014) The Incas. Wiley-Blackwell
Dresse A, Fischhendler I, Nielsen J, Zikos D (2019) Environmental peacebuilding: towards a theoretical framework. Coop Confl 54(1):99–119
Ernstorfer A, Jean I, Woodrow P, Chigas D (2016) Thinking evaluatively in peacebuilding design, implementation, and monitoring: three reflecting on peace practice (RPP) and do no harm (DNH)-infused options to strengthen the effectiveness of peacebuilding strategies and programs. Peacebuilding Evaluation Consortium. https://www.cdacollaborative.org/publication/thinking-evaluatively-peacebuilding-design-implementation-monitoring/
Evenson R, Gollin D (2003) Assessing the impact of the Green Revolution, 1960 to 2000. Science 300(5620):758–762
Fearnside P (2005) Deforestation in Brazilian Amazonia: history, rates, and consequences. Conserv Biol 19(3):680–688
Fisher J (2022) Managing environmental conflict. Columbia University Press
Fisher C (2024) Cultivating modern farms using ancient lessons, May 21. https://www.sapiens.org/archaeology/ancient-modern-farmers-climate-change/
Francopan P (2023) The earth transformed: an untold story. Knopf
Gates P (1965) Agriculture and the civil war. Knopf
Hardt, J. & Scheffran J (2019) Environmental peacebuilding and climate change: peace and conflict studies at the edge of transformation, Toda Peace Institute Policy Brief, 68. At:https://toda.org/policy-briefs-and-resources/policy-briefs/environmental-peacebuilding-and-climate-change-peace-and-conflictstudies-at-the-edge-of-transformation.html
Homer-Dixon T (1999) Environment, scarcity, and violence. Princeton University Press
Hughes J (1994) Pan's travail: environmental problems of the ancient Greeks and Romans. Johns Hopkins University Press
Ide T, Bruch C, Carius A, Conca K, Dabelko G, Matthew R, Weinthal E (2021) The past and future(s) of environmental peacebuilding. Int Aff 97(1):1–16
Intergovernmental Panel on Climate Change (2014) Climate change 2014: impacts, adaptation, and vulnerability. Cambridge University Press
Islam S, Susskind L (2012) Water diplomacy: a negotiated approach to managing complex water networks. Routledge
Kelman I (2012) Disaster diplomacy: how disasters affect peace and conflict. Routledge
Kolbert E (2014) The sixth extinction: an unnatural history. Henry Holt and Company
Krajick K (2014) Mongol Empire rode wave of mild climate, says study. Columbia Climate School, March 10. https://news.climate.columbia.edu/2014/03/10/mongol-empire-rode-wave-of-mild-climate-says-study/

Krampe F, Swain A (2021) Environmental peacebuilding. In: Richmond T (ed) The Oxford handbook of peacebuilding, statebuilding, and peace formation. Oxford University Press, pp 563–578

Laakkonen S, Vuorisalo T, Tucker R (eds) (2017) The long shadows: a global environmental history of the Second World War. Oregon State University Press

Le Billon P (2001) The political ecology of war: natural resources and armed conflicts. Polit Geogr 20(5):561–584

Mach K, Kraan C, Adger W, Buhaug H, Burke M, Fearon J, Field C, Hendrix C, Maystadt J, O'Loughlin J, Roessler P, Scheffran J, Schultz K, von Uexkull N (2019) Climate as a risk factor for armed conflict. Nature 571:193–197

Marples D (1988) The social impact of the Chernobyl disaster. Macmillan

Matthew R, Brown O, Jensen D (2009) From conflict to peacebuilding: the role of natural resources and the environment. United Nations Environment Programme

McClain S, Talla P, Bruch, C (2022) Monitoring, evaluation, and learning: approaches for building resilience and sustaining peace. https://www.ecosystemforpeace.org/compendium/monitoring-evaluation-learning-approaches-for-building-resilience-and-sustaining-peace

McNeill J (2000) Something new under the sun: an environmental history of the twentieth-century world. W.W. Norton & Company

McNeill J, Engelke P (2014) The great acceleration: an environmental history of the anthropocene since 1945. Harvard University Press

Micklin P (2007) The Aral Sea disaster. Annu Rev Earth Planet Sci 35:47–72

National Oceanographic and Atmospheric Administration (2024) A brief history of pollution. https://oceanservice.noaa.gov/education/tutorial_pollution/02history.html

Organisation for Economic Co-operation and Development (2016) Encouraging effective evaluation of conflict prevention and peacebuilding activities: toward DAC guidance. OECD Publishing. https://www.oecd-ilibrary.org/development/part-1-encouraging-effective-evaluation-of-conflict-prevention-and-peacebuilding-activities_journal_dev-v8-art37-en

Organisation for Economic Co-operation and Development (2023) Glossary of key terms in evaluation and results-based management for sustainable development (2nd edn). OECD Publishing. https://www.oecd.org/en/publications/glossary-of-key-terms-in-evaluation-and-results-based-managementfor-sustainable-development-second-edition_cfda7a8c-de-en.html

Organisation for Economic Co-operation and Development (2024) Evaluation criteria. OECD Publishing. https://www.oecd.org/en/topics/sub-issues/development-co-operation-evaluation-and-effectiveness/evaluation-criteria.html

Ostrom E (1990) Governing the commons: the evolution of institutions for collective action. Cambridge University Press

Pacific Institute (2024) Water conflict chronology. https://pacinst.org/water-conflict-chronology/

Ponting C (1991) A green history of the world: the environment and the collapse of great civilizations. Penguin Books

Radkau J (2008) Nature and power: a global history of the environment. Cambridge University Press

Ramsar Convention on Wetlands (2021) Wetlands restoration: unlocking the untapped potential of the Earth's most valuable ecosystem. https://www.ramsar.org/sites/default/files/documents/library/factsheet_wetland_restoration_general_e_0.pdf

Richards J (1993) The Mughal Empire. Cambridge University Press

Richards J (2003) The unending frontier: an environmental history of the early modern world. University of California Press

Sändig J, Dalmer N, Ide T, Vogler A (2024) From climate conflicts to environmental peacebuilding: exploring local dimensions. Environ Secur 2(1):3–20

Scheffran J, Brzoska M, Kominek J, Link P, Schilling J (2012) Climate change and violent conflict. Science 336(6083):869–871

Serrat O (1994) Deforestation in the Philippines. Unpublished manuscript, Asian Development Bank. https://www.researchgate.net/publication/291056301_Deforestation_in_the_Philippines

Serrat O (2017a) On second thought. In: Knowledge solutions: tools, methods, and approaches to drive organizational performance. Springer, pp 1101–1113

Serrat O (2017b) Identifying and sharing good practices. In: Knowledge solutions: tools, methods, and approaches to drive organizational performance. Springer, pp 843–846

Serrat O (2017c) Learning from evaluation. In: Knowledge solutions: tools, methods, and approaches to drive organizational performance. Springer, pp 879–884

Serrat O (2024a) Environmental peacebuilding: nurturing hope with analytical persuasion. Unpublished manuscript, Georgetown University. https://www.researchgate.net/publication/382637520_Environmental_Peacebuilding_Nurturing_Hope_with_Analytical_Persuasion

Serrat O (2024b) Preserving paradise: the heart of Borneo Initiative. Unpublished manuscript, Georgetown University. https://www.researchgate.net/publication/381993897_Preserving_Paradise_The_Heart_of_Borneo_Initiative

Serrat O (2024c) Safeguarding the mother of waters: environmental peacebuilding in the Mekong River Basin. Unpublished manuscript, Georgetown University. https://www.researchgate.net/publication/382520591_Safeguarding_the_Mother_of_Waters_Environmental_Peacebuilding_in_the_Mekong_River_Basin

Stearns P (2012) The industrial revolution in world history. Westview Press

Susskind L (1994) Environmental diplomacy: negotiating more effective global agreements. Oxford University Press

Swain A (2015) Water and post-conflict peacebuilding. Routledge

Swain A, Öjendal J (eds) (2018) Routledge handbook of environmental conflict and peacebuilding. Routledge

Tainter J (1988) The collapse of complex societies. Cambridge University Press

Tucker S (ed) (2005) World War I: a student encyclopedia. ABC-CLIO

Tucker M, Russell L (2004) Natural enemy, natural ally: toward an environmental history of war. Oregon State University Press

U.S. Department of Agriculture (2024) Reforestation. https://www.fs.usda.gov/managing-land/forest-management/vegetation-management/reforestation

United Nations (2024) Conflict and natural resources. https://peacekeeping.un.org/en/conflict-and-natural-resources

United Nations General Assembly (2015) Transforming our world: the 2030 agenda for sustainable development. A/RES/70/1.

Van Brabant K (2010) Peacebuilding how? Criteria to assess and evaluate peacebuilding. Interpeace. https://www.interpeace.org/wp-content/uploads/2010/08/2010_IP_Peacebuilding_How_Criteria_How_To_Assess_And_Evaluate_Peacebuilding.pdf

Vizina Y (2019) Indigenous perspectives of sustainable development. In: Filho LW (ed) Encyclopedia of sustainability in higher education. Springer

Wolf A (2007) Shared waters: conflict and cooperation. Annu Rev Environ Resou 32:241–269

World Commission on Environment and Development (1987) Our common future. Oxford University Press

Worldometer (2024) Asia population (live). https://www.worldometers.info/world-population/asia-population/

Worster D (2004) Dust Bowl: the Southern Plains in the 1930s. Oxford University Press

Young O (2002) The institutional dimensions of environmental change: fit, interplay, and scale. MIT Press

Chapter 20
Neo-Malthusian and Cornucopian Perspectives: Must It Be Either/Or?

Abstract Natural resources play a vital role in our lives but connect to conflict. This précis considers the neo-Malthusian–Cornucopian dichotomy to explore the connections between resources, conflict, and the environmental impacts of war. The précis submits that both the neo-Malthusian and Cornucopian perspectives hold valid insights: by combining their strengths with eclectic environmentalism, we can promote peace through balanced resource management and responsible growth.

20.1 Nature's Bounty, Human Strife: The Natural Resources in Our Midst

We cannot do without the air we breathe, the water we drink, or the food we eat. As de facto custodians of our planet, we should logically protect and wisely manage the natural resources in our midst. More than 30 years ago, the International Union for Conservation of Nature, the United Nations Environment Programme, and the World Wide Fund for Nature conveyed two important messages: care for the planet on which we live, and sustainability in the use of its resources (IUCN–UNEP–WWF 1991; Serrat 2017a). Alas, driven by economic and political framing, nonrenewable natural resources frequently become sources of human–human and, paradoxically, human–environment conflict as competing interests vie for access, control, and exploitation.

20.2 Natural Resources, Conflict, and the Environmental Impacts of War

Driven by the fast-changing interplay of environmental, socio-economic, and political factors, the connections between natural resources, conflict, and the environmental impacts of war are perforce multifaceted. Both scarcity and abundance of

O. Serrat, *Anthropogenic Solutions for Climate Change*,
https://doi.org/10.1007/978-981-96-8654-4_20

natural resources can lead to militarism and warfare. Where resources are scarce, conflict grows out of competition or grievance: for example, disputes over the oil fields in Kuwait, among other issues, set off the first Gulf War (1990–1991). Where resources are abundant, the so-called "resource curse" can stimulate competition and corruption as well as ethnic and political conflict: such was the case in diamond-rich countries of central and western Africa during the 1990s. Samset (2009) detected that "[t]he current scholarly debate on resource wealth and armed conflict was triggered in the late 1990s, when findings emerged of a statistical correlation between a measure for countries' natural resource endowment and the incidence of civil war" (para. 3).

Environmental factors are rarely the sole cause of conflict: however, demographic change (e.g., population growth, migration, urbanization), market pressures (e.g., increased commercialization, intensification and privatization of local economies, growing integration of national and regional economies, economic reforms), or environmental deterioration that force people to alter their livelihood strategies (e.g., altered river flows, changes in wildlife seasonal movements, droughts, floods) do not make for environmental peacebuilding. Klare (2001) surmised that "[c]onflict over valuable resources—and the power and wealth they confer—has become an increasingly prominent feature of the global landscape" (p. ix). Methodically, Weir (2020) itemized how militarism and warfare impact the environment before, during, and after conflict: the detrimental outcomes include biodiversity loss, deforestation, desertification, landscape devastation, resource depletion, soil poisoning, toxic chemical leakage, water contamination, radioactive pollution, and wildlife harm.

20.3 The Neo-Malthusian–Cornucopian Dichotomy

Named after Thomas Malthus (1766–1834), an English economist and cleric, the neo-Malthusian perspective asserts that population grows exponentially while food supply grows linearly and that both must decline when the former outpaces the latter. In contrast, the Cornucopian perspective holds an anthropocentric view of the world that emphasizes human progress and asserts that environmental problems either do not exist or can be solved through technology and the free market. (The term "cornucopian" derives from Latin *cornu copiae*, a mythical horn of plenty able to provide whatever is desired.) The adherents of Cornucopianism are sometimes called "boomsters" in contrast to "doomers" whose views are more aligned with neo-Malthusianism.

20.4 The Simon–Ehrlich Wager

Neo-Malthusian and Cornucopian worldviews on natural resources frequently refer to the Simon–Ehrlich wager in discussions of demographics change, market pressures, and environmental deterioration. Ehrlich (1968), an American biologist and environmentalist, had predicted worldwide famines due to overpopulation (and other societal upheavals) and advocated immediate action to limit population growth; responding in 1980, Simon, an American economist, challenged Ehrlich (and two other neo-Malthusian scholars) to select any raw material and a date more than a year away, betting that the commodity's price on that future date would be lower than its price at the time of the wager. Simon had rationalized that in a market economy with freely determined prices and secured property rights, a rise in the price of a valuable resource would be temporary because it would incentivize people to find more of it, use it efficiently, or develop substitutes. Despite a world population increase of over 800 million thereafter, the price of each of Ehrlich's selected metals (i.e., chromium, copper, nickel, tin, tungsten) had declined by 1990, using 1980 as an index. Ehrlich settled the wager in Simon's favor. According to Tupy and Pooley (2021), "Ehrlich and his group lost because they thought like biologists" (para. 4), who concern themselves exclusively with the carrying capacity of the environment. Then again,

> Simon won because he thought like an economist. He understood the powers of incentives and the price mechanism to overcome resource shortages. [...] He saw resource scarcity as a temporary challenge that can be solved through greater efficiency, increased supply, development of substitutes, and so on. (Tupy and Pooley 2021, para. 5)

Tupy and Pooley (2021) publicized the upcoming publication of Tupy and Pooley (2022), a paean to how population growth and innovation can promote human flourishing on an infinitely bountiful planet. If Cornucopians are right, resource wars should soon be a thing of the past: with a world population expected to surpass 10 billion by 2060 (Statista 2024), all will be for the best in the best of all possible Leibnizian worlds. But, why did population growth in earlier stages of history not bring the plenteousness that Cornucopians would have it engender? If technological advancements and market-driven solutions can be counted on to improve well-being, why are standards of living within and across countries heavily skewed, with many sizeable countries having significant proportions of their populations living below the extreme poverty line of $2.15 per person per day? And, why did 193 members states of the United Nations commit in 2015 to 17 sustainable development goals (UN General Assembly 2015)? Other aftereffects of the market-oriented calculus of the Cornucopian perspective could be mentioned, such as the privatization of resources or the imposition of austerity measures. Simon won his bet, but the state of our planet demands that we move beyond fixation on short-term price fluctuations (Maslin 2021).

20.5 Choosing Both/And Instead of Either/Or

The Simon–Ehrlich wager encapsulates the complex relationship between resource availability, population growth, and market dynamics but it cannot be that the neo-Malthusian perspective will always be right and the Cornucopian perspective will always be wrong, or vice versa. By prompting us to be mindful of finite resources and the consequences of unchecked population growth, neo-Malthusianism serves as a cautionary reminder of the need for sustainable practices. By emphasizing human ingenuity, scientific advancements, and innovations, Cornucopianism helps find new ways to meet demands and face challenges (e.g., biodegradable materials, carbon capture and storage, circular economy practices, the Green Revolution, renewable energy sources, satellite-based Earth observation). Conversely, too much focus on limitations can stifle creativity, leading to incremental changes rather than transformative ones; likewise, too much emphasis on innovations may result in impractical ideas that ignore practical constraints. Bhavnani (2009) reasoned that abundance and scarcity are two sides of the same coin. To illustrate, water is the new oil in the Gulf where, owing to economic development and a projected 30% population increase over the next decade, that region must now take steps to secure its water future (Hakimdavar 2024).

When nothing can be known with any certainty, both/and thinking embraces seeming contradictions and serves better than either/or points of view. Recognizing that differences in the distribution or accessibility of natural resources require a range of "hardware" (e.g., infrastructure, physical tools, technologies) and "software" (e.g., capacity development, collaboration platforms, conflict-sensitive programming, policy frameworks), nuanced approaches that combine neo-Malthusian and Cornucopian perspectives promote holistic understanding, informed decision-making, balanced policies, and innovation and adaptation. At personal, relational, cultural, and structural levels, eclectic environmentalism that is informed by effective questions can conduce harmony between human needs and planetary limits to avoid a vicious cycle (Serrat 2017b). Recognizing that the thinking that brought about the world's ills is not likely to get us out in the Anthropocene, Brown (2009) offered the following advice: "First we need to decide what needs to be done. Then we do it. And then we ask if it is possible" (p. xiv).

References

Bhavnani R (2009) Scarcity, abundance, and conflict: a complex new world. The Whitehead J Dipl Int Relat 10(2):65–80

Brown L (2009) Plan B 4.0: mobilizing to save civilization. Norton

Ehrlich P (1968) The population bomb. Sierra Club/Ballantine Books

Hakimdavar R (2024) Water is the new oil in the Gulf. Time, January 18. https://time.com/6556469/water-new-oil-gulf/

IUCN–UNEP–WWF (1991) Caring for the Earth: a strategy for sustainable living. Routledge

Klare M (2001) Resource wars: the new landscape of global conflict. Henry Holt

Maslin M (2021) How to save our planet: the facts. Penguin Life

Samset I (2009) Natural resource wealth, conflict, and peacebuilding. Chr. Michelsen Institute. https://www.cmi.no/publications/3283-natural-resource-wealth-conflict

Serrat O (2017a) Political economy analysis for development effectiveness. In: Knowledge solutions: tools, methods, and approaches to drive organizational performance. Springer, pp 207–222

Serrat O (2017b) Asking effective questions. In: Knowledge solutions: tools, methods, and approaches to drive organizational performance. Springer, pp 889–895

Statista (2024) Population - Statistics & facts. https://www.statista.com/topics/776/population/

Tupy M, Pooley G (2021) Resources are more abundant than ever, and people are the reason. Cato Unbound, April 21. https://www.cato-unbound.org/2021/04/21/marian-l-tupy-gale-pooley/resources-are-more-abundant-ever-people-are-reason/

Tupy M, Pooley G (2022) Superabundance: the story of population growth, innovation, and human flourishing on an infinitely bountiful planet. Cato Institute

UN General Assembly (2015) Transforming our world: the 2030 Agenda for Sustainable Development, A/RES/70/1.

Weir D (2020) How does war damage the environment? Conflict and Environment Observatory, June 4. https://ceobs.org/how-does-war-damage-the-environment/

Chapter 21
Operation Earth: Integrating Eco-Friendly Practices into the Military

Abstract If the world's militaries were a country, they would rank as the fourth-largest global emitter of carbon dioxide. This précis makes the case for greening the military; identifies opportunities in the areas of climate change mitigation but also resource management and conservation and environmental diplomacy; and signals practicable entry points in each area.

21.1 A Force for Environmental Peace

The primary mission of the military is to safeguard a nation and its citizens from external armed threats. (In some countries, the military also fosters national unity.) However, it is important to recognize that militarism and warfare do not make for environmental peace. Weir (2020) documented how the military impacts the environment before, during, and after conflict: the end results include biodiversity loss, deforestation, desertification, landscape devastation, resource depletion, soil contamination, toxic chemical leakage, water pollution, radioactive emissions, and harm to wildlife. Given the pressing issues of climate change and resource scarcity, integrating environmental conservation into military strategies and practices should be a top priority (Serrat 2024a). However, some contend that the military "does not have great tools for building things, for creating things, for bringing hope and dignity" (Matthew 2010, p. 41). Should help be required, Matthew (2010) continued, "we will send the military over [...] because we have no one else to send" (p. 48).

Does the preceding suggest that the military's primary objective is solely achieved through warfare or, at best, by preventing it? The answer is a resounding "No." Dichotomies exist precisely so they might be reconciled (Serrat 2024b). In the Anthropocene, where local, national, regional, and global environmental degradation and despoilation drive many to despair, it stands to reason that we should utilize the military's extensive logistical capabilities not only for narrow national security concerns but also for environmental peacebuilding both domestically and internationally. *De facto*, the military already undertakes additional missions beyond warfare, including peacekeeping, rescue operations, providing food and

O. Serrat, *Anthropogenic Solutions for Climate Change*,
https://doi.org/10.1007/978-981-96-8654-4_21

humanitarian assistance, offering medical aid in impoverished areas, and responding to disasters.

Efforts to optimize the military's utility should be intensified. While the fundamental purpose of the military grants it access to significant resources, it also consumes them. Consider, for example, the staggering $850 billion allocated for defense spending in the United States in 2025: that apportionment constitutes approximately 3% of the country's national income, which at an impressive $25.6 trillion (2020) is the highest globally (Hanlon 2024). According to Statista (2024), the largest armies in the world—ranked by active military personnel include China, India, the United States, Russia, and North Korea, collectively numbering around 7.5 million soldiers. Given these figures, there are ample opportunities both in the United States and other nations to promote environmental sustainability within the military. As Jens Stoltenberg, the Secretary General of the North Atlantic Treaty Organization, clarified at the high-level roundtable on *Climate, Peace, and Stability: Weathering Risk Through COP and Beyond* held in Glasgow, United Kingdom, on November 2, 2021 (as cited in European Union 2024), "We cannot choose between either green or strong armed forces; we need strong and green at the same time." The critical question is: "How can the military integrate environmental protection into its operations and policies?"

21.2 Operation Earth: Climate Change Mitigation

Determining the carbon "boot print" of the armed forces presents challenges, but estimates indicate that the military could account for up to 5.5% of global carbon dioxide (CO_2) emissions (European Union 2024). If combined, the world's militaries would rank as the fourth-largest emitter globally (European Union 2024). Not counting—if it were possible—the broader issue of the direct impact of warfare on CO_2 emissions the military has significant potential to contribute to climate change mitigation by, say:

- Decarbonizing military installations (for instance by improving the efficiency of defense estate and bases, transitioning to renewable heating, leveraging renewable power sources, and exploring carbon capture technology)[1]
- Transitioning towards greener practices in training (for instance by increasing the proportion of virtual training activities, optimizing training logistics, promoting sustainable materials and equipment, and educating military personnel about environmental impact)

[1] In Texas, at Fort Cavazos in the United States, a hybrid renewable energy project has been put into action: the initiative combines solar panels and wind turbines to supply approximately 40% of the base's energy requirements, significantly reduce CO_2 emissions, and promote energy independence. Meanwhile, the British Army is actively testing hybrid and electric vehicles, including the Electric Light Reconnaissance Vehicle: these vehicles not only reduce reliance on fossil fuels but also operate more quietly, providing a tactical advantage.

- Reducing reliance on fossil fuel-powered (and vulnerable) convoys in operations (for instance by exploring alternative transportation modes and improving route planning

21.3 Operation Earth: Resource Management and Conservation

The military can have a pivotal role in natural resource management and conservation. By incorporating environmental factors into its land use planning, training exercises, and base operations, the military can actively contribute to sustainable resource management and peacebuilding efforts. Specifically, the military can safeguard biodiversity, oversee forest management, and protect critical ecosystems by, say:

- Leveraging military readiness for conservation (for instance by investing in habitat protection and management, and collaborating with other agencies to conserve imperiled species)[2]
- Galvanizing and sharing strategic research and environmental security technology, and increasing collaboration with conservation groups
- Balancing future requirements with environmental stewardship to maintain land under military management in ways that benefit ecosystems, water resources, and wildlife (e.g., wildlife corridors)

21.4 Operation Earth: Environmental Diplomacy

In the Anthropocene, where human activities have nefarious effects on the planet and the implications of climate change continue to escalate, the military must actively participate in environmental diplomacy. This effort should be coordinated through "whole-of-government" and "whole-of-society" approaches (Serrat 2024c, d). Specifically, the military can emphasize cooperation, resilience, and global well-being by, say:

- Extending humanitarian assistance and disaster relief during environmental crises (for instance by providing aid, logistics, and medical support)

[2] As part of the Green Belt Movement in Kenya, an Environmental Soldier Program was implemented in collaboration with the Kenya Defense Forces: soldiers stationed in military cantonments worked alongside environmental conservation partners to plant trees. Similarly, the Indian Army has created "ecological battalions" with a primary focus on forest restoration to mitigate natural disasters. In Xinjiang, China, army personnel have actively promoted environmental conservation and sustainable development: efforts include restoring marginal farmland to forests and grasslands, as well as converting desert areas into arable land.

- Including environmental components in multinational exercises (e.g., disaster simulations, joint conservation efforts)[3]
- Communicating environmental initiatives (for instance by sharing success stories of sustainable practices and raising awareness among partners)
- Engaging in discussions concerning environmental protection, resource access, and territorial claims (for instance in the Arctic)
- Tasking contractors with strict environmental guidelines to minimize their impact on fragile ecosystems in war-torn countries

21.5 Transforming Military Might: Leveraging Defense Resources for Environmental Sustainability

The objectives, strategies, and operations of the military, which ranks among the world's largest polluters, are often intertwined with the upholding of extractive economies. Furthermore, substantial defense budgets divert funds that could otherwise be used to address the urgent climate crisis and the environmental degradation and despoliation our planet faces. Nevertheless, "It doesn't matter if a cat is black or white, as long as it catches mice," Chinese Communist Party chairman Deng Xiaoping (1904–1997) allegedly remarked. If we are to tackle climate change and combat environmental degradation and despoilation, we must recognize the potential role of the military in environmental peacebuilding. Key attributes of the military, relevant to the challenges we face, include its capacity to foster shared understanding, provide clear intent, build cohesive teams, exercise disciplined initiative, operate under mission orders, and accept prudent risk. In the context of environmental peacebuilding, it has become evident that we should leverage the military's abundant financial, environmental, human, infrastructure, and social capital. By deliberately integrating eco-friendly practices into the military, we have the potential to transform a major source of human–environment conflict into a formidable ally for environmental well-being. The future of our planet depends on leveraging the military's capabilities to protect and restore the environment, ensuring security not only against conflicts but also against the anthropogenic threats our civilization has imposed upon itself.

[3]After the 2004 Indian Ocean Tsunami, the U.S. military launched Operation Unified Assistance, delivering humanitarian assistance and disaster relief to stricken areas. Additionally, the biennial Rim of the Pacific Exercise, overseen by the U.S. military, features disaster response simulations and environmental awareness initiatives: these exercises encompass scenarios like oil spill responses and collaborative conservation efforts, fostering international cooperation on environmental matters.

References

European Union (2024) Greening the armies: is a sustainable approach to national defense possible? Analysis and Research Team. https://www.consilium.europa.eu/media/69640/art_greening_-armies_web.pdf

Hanlon M (2024) U.S. defense spending in historical and international context. Econofact, May 14. https://econofact.org/u-s-defense-spending-in-historical-and-international-context

Matthew R (2010) Climate change and peace. In Voices for a Culture of Peace, 34–53, Compendium of the SGI-USA Culture of Peace Distinguished Speaker Series. Culture of Peace Press.

Serrat O (2024a) ICYMI: climate security rhymes with national security. Unpublished manuscript, Georgetown University. https://www.researchgate.net/publication/378545941_ICYMI_Climate_Security_Rhymes_with_National_Security

Serrat O (2024b) Neo-Malthusian and cornucopian perspectives: must it be either/or? Unpublished manuscript, Georgetown University. https://www.researchgate.net/publication/381768461_Neo-Malthusian_and_Cornucopian_Perspectives_Must_It_Be_EitherOr

Serrat O (2024c) Beggar thy neighbor to beggar thyself: why China and the United States must cooperate to address climate change. Unpublished manuscript, Georgetown University. https://www.researchgate.net/publication/380212272_Beggar_Thy_Neighbor_to_Beggar_Thyself_Why_China_and_the_United_States_Must_Cooperate_to_Address_Climate_Change

Serrat O (2024d) Redefining norms: a modern-day Marshall plan for climate adaptation and climate justice. Unpublished manuscript, Georgetown University. https://www.researchgate.net/publication/380530326_Redefining_Norms_A_Modern-Day_Marshall_Plan_for_Climate_Adaptation_and_Climate_Justice

Statista (2024) Largest armies in the world ranked by active military personnel as of January 2024. https://www.statista.com/statistics/264443/the-worlds-largest-armies-based-on-active-force-level/

Weir D (2020) How does war damage the environment? Conflict and Environment Observatory, June 4. https://ceobs.org/how-does-war-damage-the-environment/

References

Chapter 22
Preserving Paradise: The Heart of Borneo Initiative

Abstract More timber was traded from Borneo in the last two decades of the twentieth century than from the combined Amazon and Congo regions and this précis characterizes the destruction of Borneo's rainforest as self-inflicted violence, encouraged by consumerism. From that optic, the précis showcases the Heart of Borneo Initiative that was launched in 2007 as environmental peacebuilding in a grounded context from which much can—and should—be learned. The Heart of Borneo Initiative is a conscious effort to integrate social and environmental issues, cultivate cooperation rather than competition, and manage a resource or commons collaboratively.

22.1 Borneo: Paradise Under Siege

Surpassed in size only by Greenland and New Guinea, the Island of Borneo is the third largest island in the world with an area of 748,168 square kilometers (288,869 square miles) (Britannica n.d.). Situated at the geographic center of maritime Southeast Asia, Borneo is mostly covered by a dense tropical rainforest, estimated to be around 140 million years old, that together with other equatorial forests functions as the "lungs of the earth" (Britannica n.d.). The floral and faunal populations of Borneo are extremely varied: they count 222 mammals (including 44 endemic, meaning, they are not found anywhere else), 420 birds (37 endemic), 100 amphibians, 394 fish (19 endemic), and approximately 15,000 plants (about 5000 of which are endemic)—including such distinct species as the Bornean orangutan, the proboscis monkey, the Bornean pygmy elephant, the Bornean clouded leopard, the Sumatran rhinoceros (also known as the Bornean rhinoceros), the giant pitcher plant, and the Rafflesia flower, the largest on Earth (World Wide Fund for Nature n.d.). Borneo's forests provides essential ecosystem services including carbon sequestration, of paramount importance in our age of climate change; soil

The original version of the chapter has been revised. A correction to this chapter can be found at https://doi.org/10.1007/978-981-96-8654-4_27

O. Serrat, *Anthropogenic Solutions for Climate Change*,
https://doi.org/10.1007/978-981-96-8654-4_22

stabilization; and water regulation. Originating from the central highlands, Borneo's navigable rivers are vital for transportation and freshwater supply. Additionally, Borneo's natural resources support the livelihoods of indigenous communities such as the Dayak who rely entirely on the forests for food, medicine, and cultural practices. Borneo is shared by Brunei Darussalam, Indonesia, and Malaysia: approximately 73% of the island is Indonesian territory and the East Malaysian states of Sabah and Sarawak make up about 26% of the island (Britannica n.d.). Borneo is home to about 23 million people, of which more than two-thirds are in Indonesian territory, with the majority living in costal lowlands and cities (Britannica n.d.). An estimated 12 million local and indigenous peoples depend on the Heart of Borneo not just to clean the air by drawing in carbon dioxide from the atmosphere and breathing out oxygen but also, more immediately, for food, livelihood, and medicine.

Borneo's natural resources are in dire straits, at a time when our planet seems more fragile than it ever has been. In the early 1900s almost all of Borneo was draped in untouched, wild, and vast forests: from the 1960s and especially in the 1980s and 1990s, these forests were burned, logged, cleared, and converted under concessions at a rate unprecedented in history, with more timber cut and traded from Borneo in the last two decades of the twentieth century than from the combined Amazon and Congo regions (World Wide Fund for Nature 2010). Industrial logging and the expansion of palm oil plantations in Sabah, Sarawak, and Kalimantan, the name of the Indonesian portion of the island, were the main drivers of environmental damage in Borneo, with attendant illegal logging and slash-and-burn land clearing. In 2022, countries such as Australia, China, Japan, Malaysia, the Netherlands, the United Kingdom, and the United States were principal destinations of wood products (i.e., building materials, fancy furniture) from Borneo, with the United States and Japan easily topping the list; Malaysia, which incorporates the states of Sabah and Sarawak, was rather puzzlingly the fifth largest importer of Indonesian wood products (Observatory of Economic Complexity 2024). Once cleared, the forests were turned over to palm oil (but also pulpwood and timber) plantations. Indonesia and Malaysia are, respectively, the first and second largest producers of palm oil, which is found in nearly every household item from bread to chocolate to lipstick to shampoo, not mentioning industrial applications and bioenergy. Together, Indonesia and Malaysia accounted for about 85% of global supply in 2021, or about 63% and 22% respectively (Ritchie 2021). Potential substitutes for palm oil include canola and sunflower seed oils, heterotrophic algal oil, and peanut and rice bran oils but finding alternatives that match palm oil's affordability remains a challenge; for this reason, consumer choices play a crucial role in protecting Borneo's forests.

The disruption of ecosystem services in Borneo was characterized by reduced carbon sequestration from the destruction of forests and the drainage of carbon-rich peatland for agricultural expansion and infrastructure development; loss of biodiversity from the disappearance of critical habitats impaired water regulation causing increased flooding during the rainy season and droughts during the dry season; soil erosion resulting in loss of fertile land and the sedimentation of rivers, in turn affecting water quality and aquatic life; and the displacement of indigenous communities. Fires ignited to clear land for crops sometimes spin out of control in dry El Niño seasons, for example in 1991, 1994, 1997, 2006, 2015, 2019, and 2023, but also spread on highly flammable peatlands that had been turned over to rice production, thereby contributing to the world's greenhouse gas emissions (Jong 2023). There are no confirmed reports of any species that have gone completely extinct in recent decades but the Bornean orangutan and the pygmy elephant are at high risk and the number of Sumatran rhinoceroses is critically low (World Wide Fund for Nature 2007a).

The deforestation of Borneo continued through the 2000s. Still, using a data set spanning nearly two decades, researchers from the Center for International Forestry Research found that the expansion of oil palm plantations in Borneo had slowed from 2012 (United Nations Environment Programme 2019). The slowing trend in deforestation owes to declining crude palm oil prices and more effective government regulations but continuous efforts and conservation programs are assuredly needed. Rainforest means rain: McAlpine et al. (2018) found that large-scale deforestation across Borneo has led to higher temperatures and less precipitation over the past 60 years, with likely stark consequences not only for Borneo's remaining forests and wildlife but also its people and its agricultural sector. Box 22.1 recollects the rich biodiversity of the Kinabatangan River, a vital ecological corridor in Borneo.

Box 22.1 Echoes of Eden: Celebrating the biodiversity of Kinabatangan River

- It is the year 2018. I am travelling from Sandakan, in the Malaysian state of Sabah, to the lower reaches of the Kinabatangan River. From the mountains in Tongod District, the river snakes through 560 kilometers (350 miles) before surrendering to the Sulu Sea. To reach our lodge, we drive for two and a half hours across the monocrop plantations of oil palm that claim sovereignty over one quarter of Sabah. The plantations encroach right up to the river, leaving a thin corridor of rainforest for local fauna and flora. The river features riparian forests, nipah swamps, and oxbow lakes. The riverbanks are a sanctuary for Borneo's iconic wildlife: orangutans (that is, "persons of the forest"), pygmy elephants, bearded pigs, proboscis monkeys, clouded leopards, silver langurs, long-tailed macaques, hornbills, tarsiers, Storm's storks … There are bloodthirsty leeches too. The canopies above the river remind us of the delicate balance of life. On a tall tree, a proboscis monkey with a uniquely enlarged nose holds off challengers: one of them falls to the river and swims with webbed hands and feet toward the other bank, in seemingly chastised shame. The victorious alpha male honks, shakes the tree, and celebrates one more day at the apex of his life. Efforts to preserve the river's corridor of life are ongoing, with conservation initiatives working alongside local communities to restore and reconnect fragmented habitats. In this corner of our globalizing world, thankfully protected from the deforestation that has razed Borneo, every turn of the river tells a story of diversity and coexistence.

22.2 The Rationale for Peacebuilding

The United Nations Environment Programme (2019) estimated that only half of Borneo's forest cover remains, down from 75% in the mid-1980s. The destruction of Borneo's forests is self-inflicted violence, encouraged by consumerism overseas, the consequences of which also extend beyond the island on account of reduced carbon sequestration, thence climate change. Governments are responsible for domestic governance and policy-making as well as foreign affairs: and so, the governments of Indonesia and Malaysia must find environmental pathways to peace. By integrating technical, restorative, and sustainable trajectories for environmental peacebuilding, Malaysia and Indonesia can create a more collaborative and stable environment in Borneo (Dresse et al. 2019). In Borneo, environmental peacebuilding involving technical, restorative, and sustainable trajectories can be a context-sensitive, win–win approach for several reasons:

- **Addressing Root Causes.** Peacebuilding shines a light on the underlying causes of deforestation and environmental degradation: by tackling related issues, it reduces the pressure on forests and promotes more sustainable land use practices.
- **Building Trust.** Peacebuilding fosters trust between different groups, which is essential for long-term cooperation and human–environment conflict resolution. Trust can help to promote inclusive dialogue and prevent destructive activities by encouraging transparency and accountability.
- **Fostering Cooperation.** Peacebuilding initiatives encourage dialogue and cooperation between stakeholders (e.g., academia and scientists, central government, civil society, donors, local communities, local government, local parliament, neighboring countries, private companies and investors, state-owned enterprises) to collectively exert coercive and normative pressure that advances sustainable practices and biodiversity conservation.
- **Empowering Local Communities, Including Indigenous Peoples.** Involving local communities, including indigenous peoples, in peacebuilding efforts ensures that their voices are heard and their rights are respected, which can lead to better forest management and conservation practices as communities are more likely to protect the forests they depend on.
- **Promoting Sustainable Development.** Peacebuilding aligns with sustainable development goals by promoting economic development that does not come at the expense of the environment.

22.3 The Heart of Borneo Initiative: Peacebuilding at the Local Level

Crises are opportunities for change. In 2007, the governments of Brunei Darussalam, Indonesia, and Malaysia committed to the Heart of Borneo Initiative in 2007 under a joint Declaration (World Wide Fund for Nature 2007b). The Heart of Borneo Initiative is a unique conservation program catalyzed by the World Wide Fund for Nature (WWF) to protect about 220,000 square kilometers (84,940 square miles) of forested region in Borneo, which constitutes nearly one-third of the entire island. Kalimantan, Sabah and Sarawak, and Brunei Darussalam account for about 60%, 38%, and 2%, respectively, of the area given over the Heart of Borneo Initiative. The Heart of Borneo Initiative aims to ensure effective management of resources and conservation of a network of protected areas, productive forests, and other sustainable land uses through trinational, rolling, government- and donor-funded, five-yearly national strategic plans of action and related interventions, initially across five main programs:

- **Transboundary Collaboration.** The program's objective is to address issues of management of natural resources and socio-economic welfare of local people on the border areas. Here, collaboration implies shared ownership and interest in a specific outcome.
- **Protected Area Management.** The program's objective is to enhance and promote effective management of protected areas within the Heart of Borneo Initiative area, with the emphasis at those situated on the common border, in order to conserve and maintain forest biodiversity and the ecological linkages.
- **Sustainable Natural Resources Management.** The program's objective is to manage the natural resources outside the protected areas network through the development and implementation of sustainable land uses.
- **Ecotourism Development.** The program's objective is to recognize and protect the value of special natural and cultural places or sites within the Heart of Borneo Initiative area.
- **Capacity Building.** The program objective is to ensure the effective implementation of the Heart of Borneo Initiative at all levels, both in the public and private sectors and at local community levels (World Wide Fund for Nature 2009).

Inspired by engagement and voluntarism, the Heart of Borneo Initiative is a conscious collaborative effort to cultivate cooperation and collective action for successful management of a resource or commons, aiming to prove that over-exploitation of natural resources is not inevitable (Hardin 1968; Ostrom 1990). Overall, the Heart of Borneo Initiative exemplifies how environmental peacebuilding can be a powerful tool for conservation by promoting cooperation, respecting local cultures, and supporting sustainable development. Vitally, given the scale and scope of its programs and associated actions across three countries, the Heart of Borneo Initiative offers plentiful opportunities to acclimatize environmental peacebuilding, including its principles and practices, in a grounded local context. In 2013, Brunei

Darussalam, Indonesia, and Malaysia commemorated the establishment of the Heart of Borneo Initiative, reaffirmed their commitment, and emphasized a green economy approach to environmental conservation. The Heart of Borneo Initiative's partners in conservation include the Asian Development Bank, the Global Environment Facility, the United Nations Development Programme, United Nations Environment Programme, and the World Wide Fund for Nature, with bilateral contributions from Australia, Denmark, Germany, Japan, the Netherlands, Singapore, the United Kingdom, the United States, and other donors. The Heart of Borneo Declaration has been endorsed by regional and international agreements such as the Asia-Pacific Economic Cooperation (APEC) forum; the Association of South East Asian Nations (ASEAN); the Brunei-Darussalam–Indonesia–Malaysia–Philippines East Asean Growth Area (BIMP–EAGA); and the United Nations Convention on Biological Diversity (UNCBD).

22.4 Sustaining the Pulse of Borneo

This précis on environmental peacebuilding at the local level is not the place to evaluate accomplishments and shortcomings across the several national strategic plans of action that the governments of Brunei Darussalam, Indonesia, and Malaysia have formulated since 2008. Numerous projects calling for dedicated and holistic attention have been implemented across the landscape of the Heart of Borneo Initiative, contributing to the preservation of a unique ecosystem. Of course, the Heart of Borneo Initiative continues to face challenges owing to agricultural pressures, economic development at border areas, forest fires, fulfilment of subsistence needs, land ownership and property rights, over-hunting of wildlife, proposals for road building infrastructure, the timber industry, and the need for sustainable funding. Concerns have been raised about the ongoing relocation of Indonesia's capital from Jakarta to Nusantara in Eastern Borno, about 200 kilometers (125 miles) away from the Heart of Borneo, even if this is an area that had in earlier times been cleared for logging, coal mining, and oil palm plantations. The threats to the Heart of Borneo will not disappear overnight. Necessarily, the Heart of Borneo Initiative is a long-term endeavor that underscores the need for effective regulation, ongoing vigilance, and commitment to sustainable practices. Notwithstanding, what progress was made within a few years of launching the Heart of Borneo Initiative, for example in conserving species, working with people, expanding sustainable forests, transforming markets, and innovative conservation financing, have underscored the power of environmental peacebuilding, particularly collaboration and cooperation, in preserving our planet's precious ecosystems (World Wide Fund for Nature 2012). By valuing natural capital, engaging in partnerships, and implementing strategic actions, the Heart of Borneo Initiative strives to secure a better future for this vital region.

References

Britannica (n.d.) Borneo island, Pacific Ocean. https://www.britannica.com/place/Borneo-island-Pacific-Ocean
Dresse A, Fischhendler I, Østergaard Nielsen J, Zikos D (2019) Environmental peacebuilding: towards a theoretical framework. Coop Confl 54(1):99–119
Hardin G (1968) The tragedy of the commons. Science 162(3859):1243–1248
Jong H (2023) El Niño leads to more fires and toxic air pollution in Indonesia. Mongabay, September 15. https://news.mongabay.com/2023/09/el-nino-leads-to-more-fires-and-toxic-air-pollution-in-Indonesia/
McAlpine C, Johnson A, Salazar A, Syktus J, Wilson K, Meijaard E, Seabrook L, Dargusch P, Nordin H, Sheil D (2018) Forest loss and Borneo's climate. Environ Res Lett 13(4):1–10
Observatory of Economic Complexity (2024) Wood products in Indonesia. https://oec.world/en/profile/bilateral-product/wood-products/reporter/idn
Ostrom E (1990) Governing the commons: the evolution of institutions for collective action. Cambridge University Press
Ritchie H (2021) Palm oil. Our World in Data. https://ourworldindata.org/palm-oil
United Nations Environment Programme (2019) Deforestation in Borneo is slowing, but regulation remains key, February 18. https://www.unep.org/news-and-stories/story/deforestation-borneo-slowing-regulation-remains-key
World Wide Fund for Nature (2007a) WWF captures extraordinary video of rare Borneo rhino, April 24. https://wwf.panda.org/wwf_news/?99560/WWF-captures-extraordinaryvideo-of-rare-Borneo-rhino
World Wide Fund for Nature (2007b) Declaration on the heart of Borneo initiative: three countries, one conservation vision, February 12. https://wwf.panda.org/discover/knowledge_hub/where_we_work/borneo_forests/about_borneo_forests/declaration/
World Wide Fund for Nature (2010) Heart of Borneo: background info for media. https://wwfasia.awsassets.panda.org/downloads/hob_media_background_briefing_faqs.pdf
World Wide Fund for Nature (2012) The heart of Borneo beats strong. https://www.worldwildlife.org/stories/the-heart-of-borneo-beats-strong
World Wide Fund for Nature. (2009). Heart of Borneo Strategic Plan of Action. https://wwf.panda.org/discover/knowledge_hub/where_we_work/borneo_forests/about_borneo_forests/hobstrategy/
World Wide Fund for Nature (n.d.) Borneo wildlife. https://wwf.panda.org/discover/knowledge_hub/where_we_work/borneo_forests/about_borneo_forests/borneo_animals/

Chapter 23
Safeguarding the Mother of Waters: Environmental Peacebuilding in the Mekong River Basin

Abstract The Mekong River weaves together cultures, sustains biodiversity, and underpins livelihoods; however, competing interests imperil its delicate equilibrium. This précis traces the evolution of environmental peacebuilding in the Mekong River Basin and offers vital recommendations to protect the Mother of Waters.

23.1 The Names of the Mekong

What's in a name? Stretching nearly 4350 kilometers (2700 miles) from the Tibetan plateau down to the South China Sea, the Mekong River flows through six countries and their diverse landscapes: each country has its own name for it, highlighting the river's vital role in the lives and traditions of the people who live along its banks (Britannica 2024). In China, the Mekong River is known as Lancang Jiang, meaning "Turbulent River." In Laos and Thailand, it is called Mae Nam Kong (or Mae Nam Khing in Myanmar), which translates to "Mother of Waters." In Cambodia, the river is referred to as Tonle Thom or "Great River," while in Vietnam it is known as Cuu Long, that is, "Nine Dragons," representing the nine branches of the river in its delta (Britannica 2024). As its names suggest, *pace* Juliet's declamation on labels and conventions in Shakespeare's (1993) *Romeo and Juliet* (Act 2, Scene 2, lines 38–49), the Mekong River is not just water: it is a lifeline for 60 million people, providing food, water, and livelihoods as well as a critical means of transportation and trade (Britannica 2024). The Mekong River supports the largest inland fisheries in the world, notably in the Tonle Sap ("Great Lake") of Cambodia that it nourishes during the monsoon season (May to October), accounting for approximately 80% of protein consumption and up to 25% of the global freshwater catch (World Wide Fund for Nature 2024). In addition, the fertile lands that the Mekong River irrigates are ideal for agriculture, earning the Mekong River Delta the nickname of "Asia's rice bowl" (Britannica 2024). Beyond its economic importance, the Mekong River is home to a diverse ecosystem, including at least 1200 freshwater species but also endangered species such as the Mekong giant catfish, the Irrawaddy dolphin, and the giant freshwater stingray (World Wide Fund for Nature 2024).

O. Serrat, *Anthropogenic Solutions for Climate Change*,
https://doi.org/10.1007/978-981-96-8654-4_23

23.2 Powering Riparian Tensions

The transnational region of the Mekong River, aka the Greater Mekong Subregion, is home to more than 300 million people. Predictably, perhaps, there have been tensions between riparian states over the use of the Mekong River and its tributaries. Intense debate about China's dams on the Mekong River has been raging for a long time. From its downstream perspective, Vietnam has charged that China's dams, exacerbating years of climate change in a region that is particularly vulnerable to that phenomenon, contributed in 2016 to the worst drought in 90 years, and also damaged its rice crops in 2019 (Goo 2021). Laos has shared similar concerns about the impact of Chinese dams on its farming and fishing industries, but has at the same time advertised plans to become the "Battery of Southeast Asia," with existing and possible dams on tributaries of the Mekong River and two new dams on the mainstream (Goo 2021). However cost effective it may be, why would Thailand, Cambodia, and Vietnam wish to buy electricity from Laos if its generation in that upstream riparian state has negative environmental and social consequences for their populations? Would they not seek retribution and retaliate somehow? But, in truth, nearly every tributary in every country of the Greater Mekong Subregion is now blocked by a dam for hydropower generation, flood control, irrigation, water storage, or mixed use (including navigation and recreation), posing a significant threat to the annual flood dynamics that drive fish migration and downstream nutrient-rich sediment distribution for flood-recession agriculture. Eyler and Kwan (2024) estimated that 745 dams are complete or under construction on the mainstream and tributaries of the Mekong River, "rendering the once connected, natural river system into a highly engineered watershed utilized for hydropower production, irrigation, urban water supply, and flood control" (para. 2).

23.3 Unraveling the Mekong River's Ecological Fabric

While dams cast a long shadow over the Mekong River and its tributaries, the Greater Mekong Subregion's environmental woes extend beyond concrete barriers. What with economic development, population growth, urbanization, and enlarged consumption and production patterns, the Greater Mekong Subregion has also been impacted by road construction, agricultural deforestation, mining, overfishing, palm oil and rubber plantations, logging and illegal timber trade, as well as illegal wildlife trade (World Wide Fund for Nature 2024). In the 1970s, the Greater Mekong Subregion was one of the world's most densely forested areas: deforestation has since reduced its forest cover by a third and the subregion is on a trajectory to lose another third by 2030 (Carroll 2018). The loss of flooded forests, in particular, has impacted the aquatic food chain, putting further pressures on biodiversity, the fishing industry, and the 40 million people who take part in fishing-related activities in the Greater Mekong Subregion. The loss of the flooded forests of the Tonle

Sap—one of the most productive and biodiverse wetlands in the world, often described as the "beating heart" of the Mekong River—is a case in point.

23.4 The Case for Environmental Peacebuilding

Environmental peacebuilding is an approach that leverages the sustainable management of natural resources to prevent conflict and build peace: it emphasizes the potential for environmental cooperation to support peace and stability, especially in regions where resources are shared across borders (Krampe and Swain 2021). The concept and practice of environmental peacebuilding rests on the idea that the biophysical environment's inherent characteristics can act as incentives for cooperation and peace, rather than competition and conflict (Dresse et al. 2019). Transboundary cooperation is especially relevant to environmental peacebuilding because many environmental challenges (e.g., loss of biodiversity, climate change, pollution, water exploitation or scarcity) transcend political boundaries and summon us to think over longer time frames (Ide et al. 2021). By working together on shared environmental issues, neighboring countries can build trust, foster dialogue, and create institutions that promote further integration and conflict resolution (Ide et al. 2021). Peace parks, to give one example, exemplify ecological diplomacy by creating protected areas that cross international borders, thereby promoting biodiversity conservation and fostering cooperation between neighboring countries (Hutagalung 2024). In the Mekong River Basin, environmental peacebuilding to safeguard the "Mother of Waters" would seem eminently desirable in view of the sociopolitical challenges of fast-paced economic development, including competing national interests and the influence of major stakeholders such as China. Not to forget, environmental peacebuilding aligns seamlessly with the Sustainable Development Goals that the United Nations established in 2015 (United Nations General Assembly 2015).

23.5 Environmental Peacebuilding in the Mekong River Basin

The Mekong River Basin is divided into upper and lower regions, each with its distinctive character. The Mekong River's headwaters begin high in the Tibetan plateau, considered to be the Upper Mekong region that includes China. The Lower Mekong region is where about three-fourths of the drainage area of the Mekong lies within Laos, Thailand, Cambodia, and Vietnam. In 1952, the Economic Commission for Asia and the Far East (ECAFE) of the United Nations conducted a study, in cooperation with the four lower riparian states, that highlighted the immense potential of the Mekong River for hydroelectric power and irrigation development. Recognized that successful projects—especially those crossing national boundaries—require joint planning and data sharing, ECAFE's report emphasized the need for close inter-riparian cooperation and coordination. To promote peace, regional cooperation, and development in the Lower Mekong region, ECAFE then encouraged the creation of an intergovernmental agency for joint water management. In 1957, the Mekong Committee was created and tasked to promote, coordinate, supervise, and control water resources development projects in the Lower Mekong region. After an initial period of enthusiasm, however, momentum subsided in the 1970s because of political turmoil, instability, and war. Nevertheless, the Mekong Committee's early institutional architecture provided a forum for dialogue that was sustained even in times of regional hostilities and laid the groundwork for the Mekong River's governance in subsequent peaceful times of rapid economic development. Notably, China was not part of this early institutional architecture due to its non-membership in the United Nations at the time.

In 1995, Laos, Thailand, Cambodia, and Vietnam signed the Agreement on the Cooperation for the Sustainable Development of the Mekong River Basin (herein referred to as the 1995 Mekong Agreement), a landmark treaty aimed at promoting sustainable development and cooperation. The Mekong Agreement established the goals, objectives, and underlying principles by which the four countries intend to cooperate, summarized as follows:

- To cooperate in all fields of sustainable development, use, management and conservation of water and related resources of the Mekong River Basin, in a manner that optimizes multiple uses and benefits of water and related resources in the Mekong River Basin including irrigation, hydropower, navigation, flood control, fisheries, timber floating, recreation, and tourism (Article 1).
- To promote the sustainable development of the basin's full potential and prevent wasteful use, with an emphasis on joint and/or basin-wide development projects and programs (Article 2).
- To protect the environment, natural resources, aquatic life and conditions, and ecological balance of the Mekong River Basin and minimize pollution and other harmful effects (Articles 3 and 7–10).
- To utilize the waters of the Mekong River system in a reasonable and equitable manner (Articles 4–6) (Mekong River Commission 2017, para. 1).

The Mekong Agreement established the Mekong River Commission, an intergovernmental agency, "[t]o promote and coordinate sustainable management and development of water and related resources for the countries' mutual benefit and the people's well-being" (Mekong River Commission 2024). "An economically prosperous, socially just, environmentally sound, and climate resilient Mekong River Basin" (Mekong River Commission 2024), the vision of the Mekong River Commission, embodies a balance between economic development (SDGs 2, 7, 8, 9), social justice (SDGs 1, 2, 5, 6, 17), and environmental integrity (SDGs 6, 15), with climate resilience (SDGs 1, 13) as a cross-cutting focus (United Nations General Assembly 2015). SDG 16, which focuses on peace, justice, and strong institutions, serves as a critical bridge between environmental protection and broader development objectives (United Nations General Assembly 2015).

China and Myanmar remained outside the framework but later decided to participate as dialogue partners, sharing full-year data on water flow to provide a comprehensive picture. There are now calls for the sharing of information on sediment load across the upper and lower regions of the Mekong River Basin. Certainly, China's reluctance to join the Mekong River Commission has complicated transboundary cooperation. An uncooperative hydro-hegemon, China built dams upstream without consulting its neighbors. That said, China now seems to embrace active water diplomacy. The Lancang–Mekong Cooperation format that was introduced in 2016—and includes all six riparian states—suggests China's belated acceptance that environmental peacebuilding needs more than dams; it needs bridges of understanding (http://www.lmcchina.org/eng/). Lancang–Mekong Cooperation targets areas of cooperation having to do with water resources, trade and investment, connectivity, agriculture and poverty reduction, environmental protection, and people-to-people exchanges (http://www.lmcchina.org/eng/).

In April 2023, Laos, Thailand, Cambodia, and Vietnam issued the Vientiane Declaration, in which they reaffirmed their commitment to cooperation in the Mekong River Basin, acknowledged both the opportunities and adverse impacts of development on the river, and called for innovative solutions to help address the many challenges the river faces (United Nations Environment Programme n.d.). The Vientiane Declaration reemphasized the role of the Mekong River Commission as a regional knowledge hub, with the website of the Mekong River Commission providing public access to data for river monitoring and forecasting (https://www.mrcmekong.org/).

23.6 The Challenges of Transboundary Environmental Agreements

Despite riparian tensions arising from rapid modernization, poverty, and water utilization, the Mekong River Commission has maintained dialogue: perforce, it must acknowledge national interests whilst endeavoring to safeguard the Mekong River's health. The Mekong River Commission is not a regulatory body, a regime that has advantages but also disadvantages: it keeps countries at the table without policing, lets the Mekong River Commission add independent advice and recommendations, but does not grant the commission the authority to enforce decisions, leading to accusations that the riparian states use the Mekong River Commission as a shield to deflect criticism for their actions. With the assistance of donors (e.g., the Asian Development Bank, the Global Environment Facility, the World Bank, Australia, the European Union and its member countries, the United States, Japan), the Mekong River Commission has introduced tools (e.g., the Mekong Water Utilization Project, the Mekong Integrated Water Resources Management Project, the Basin Development Plan) to help build in Laos, Thailand, Cambodia, and Vietnam the capacity to address regional politics and development imperatives. By so doing, the

Mekong River Commission has contributed valuable input to the governance of the Mekong River Basin. Notwithstanding, the Mekong River Commission and its tools have not provided a final solution for resource exploitation. Large-scale threats to the Mekong River Basin's ecosystem persist or worsen, chief among which are dam development, climate change, habitat loss from land use change and deforestation, and illegal wildlife trade (World Wide Fund for Nature 2024). If they are to face shared challenges in a shared future, the riparian states must propel environmental peacebuilding in the Mekong River Basin to a higher level.

Sorry to say, the travails of the Mekong River Commission are par for the course. Implementing transboundary environmental agreements is very challenging:

- **Political Tensions.** Historical conflicts and ongoing political tensions can hinder cooperative efforts and trust-building among nations. In the case of shared water resources, upstream states are wont to argue for their right to use a river as they see fit while downstream states bear the brunt of environmental impacts from upstream actions.
- **Economic and Social Disparities.** Economic inequalities and social distinctions between countries often result in conflicting priorities and challenges related to resource allocation.
- **Legal and Institutional Disparities.** Different countries operate under diverse legal frameworks and possess varying institutional capacities. These differences can complicate the harmonization of policies and the effective enforcement of regulations.
- **Data and Information Sharing.** Effective cooperation relies on transparent and consistent data sharing. However, differences in data collection methods, standards, and the willingness to share sensitive information pose challenges.
- **Funding and Resources.** Adequate funding and resources are indispensable for implementing and monitoring agreements. Limited financial resources can impede progress.
- **Public Participation and Awareness.** Engaging local communities and stakeholders is critical for the success of transboundary agreements. A lack of public awareness and participation can undermine implementation efforts.
- **Climate Change.** The impacts of climate change—such as altered water availability and more frequent extreme weather events—introduce complexity into managing shared resources. Adaptive and flexible agreements are essential.

23.7 Transforming Environmental Cooperation into Political Cooperation

Paraphrasing Diamond (2005), who pondered why some societies make disastrous decisions, the issue is not that decision-makers in the riparian states fail to anticipate a problem before it arrives, fail to perceive it when it does arrive, or fail to even try to solve it; more correctly, it is that they try to solve the problem but fail to succeed.

Of course, the work of environmental peacebuilding is complicated by the sheer number of relevant ecological, economic, and political variables. Notwithstanding, give the above considerations, addressing the challenges that face the Mekong River Basin requires stronger political will, more effective communication, and firmer commitment to collaborative problem-solving. In a word, the riparian states must immerse themselves in successful decision-making (Serrat 2017a, b).

Ecological interdependence demands cooperative action. In such cases, Carius (2008) discerned that:

> The majority of ecological peace initiatives can be classified in one of three partly overlapping categories: (a) initiatives to prevent conflicts that are directly related to the environment; (b) attempts to initiate and sustain a dialogue on transboundary environmental cooperation between parties to a conflict; and (c) initiatives that are directed at achieving lasting peace by promoting conditions for sustainable development. (Carius 2008, p. 6)

Environmental peacebuilding in the Mekong River Basin falls under Carius's (2008) third category. Accepting that the complexity of a problem—and connected need for sustained political commitment—can hamper transboundary environmental cooperation, initiatives that aim to at achieve lasting peace require a long-term perspective, participation by local and nongovernmental organizations, and the development of administrative, economic, and social capacities for action that transcends polarization (Carius 2008). A commission is appointed, usually on a temporary basis, to carry out specific tasks or investigations. An authority is an organization or body that, buttressed by enabling acts granting it legitimacy, has the power to make decisions and enforce rules in a specific field or area (Serrat 2023). Taking a leaf from Carius (2008), the riparian states should realize apropos the Mekong River Commission's mandate that if technical cooperation is characteristically initiated in a "non-politicized" atmosphere, true impact can only happen when we shift gears from technical discussions to genuine political collaboration (Serrat 2004). Institutions are the *sine qua non* of environmental and political cooperation: therefore, the Mekong River Commission should be redesigned so that—in addition to serving as a regional knowledge hub—it might directly contribute to the development of a political agenda for active cooperation based on common understanding. Capacitating and empowering the Mekong River Commission would respond to the Vientiane Declaration's (otherwise perfunctory) call for innovation and cooperation for a water secure and sustainable Mekong River Basin.

23.8 A Future Agenda for Environmental Peacebuilding in the Mekong River Basin

Under a collaborative project on the future of environmental peacebuilding, Brown and Nicolucci-Altman (2022) invited actors and players to promote inclusivity and community-based approaches in environmental peacebuilding, advocate for supportive leadership, weave peacebuilding into policies, implement binding international frameworks, address environmental tensions proactively, build and share evidence, and foster collaboration for a peace-positive and nature-positive approach. Brown and Nicolucci-Altman (2022) somehow reconciled the views of five separate institutions—no mean feat—to submit a future agenda for environmental peacebuilding, relevant also to the Mekong River Basin. That said, Brown and Nicolucci-Altman's (2022) agenda should also incorporate the following concerns:

- **Long-Term Sustainability.** Emphasize the need for long-term sustainability by fitting in environmental peacebuilding into national development plans and ensuring continuous funding and support.
- **Policy Coherence.** Advocate for policy coherence across different levels of governance to ensure that environmental peacebuilding efforts are aligned with broader development and security policies.
- **Adaptive Management.** Promote adaptive management practices that allow for flexibility and responsiveness to changing environmental and social conditions.
- **Capacity Building.** Highlight the importance of building local capacities through training and education programs to empower communities to manage their resources sustainably and resolve conflicts independently.
- **Private Sector Engagement.** Discuss strategies for engaging the private sector more effectively, including incentives for businesses to adopt sustainable practices and contribute to peacebuilding efforts.
- **Technological Integration.** Emphasize the role of technology in enhancing environmental peacebuilding efforts, such as using satellite imagery for monitoring, data analytics for predicting conflicts, and blockchain for transparent resource management.
- **Gender and Social Inclusion.** Strengthen the focus on gender and social inclusion by developing specific strategies to involve women, youth, and marginalized groups as active participants and leaders in environmental peacebuilding.
- **Public Awareness Campaigns.** Implement public awareness campaigns to educate and engage the broader public on the importance of environmental peacebuilding and how they can contribute.
- **Cross-Sector Collaboration.** Encourage collaboration between different sectors, such as environmental organizations, peacebuilding institutions, and development agencies, to create a more cohesive approach.

A future agenda for environmental peacebuilding in the Mekong River Basin would do well to heed Brown and Nicolucci-Altman's (2022) suggestions and the nine real-world concerns listed above. But someone still needs to step up to the plate and take charge. Looking back at 25 years of the Mekong River Commission, Haefner (2020) lamented that "[the Commission] is seen by many as ineffectual, sidelined by its own member states' insistence on absolute national sovereignty, and marked by mistrust and miscommunication" (para. 5). Accordingly, this précis suggests that—30 years after it was established to achieve the vision of "[a]n economically prosperous, socially just, and environmentally sound Mekong River Basin" (Mekong River Commission 2024)—the Mekong River Commission should be overhauled to thrive, not struggle along by design. By the Mekong River Commission's (2021) admission, the outlook for water flow conditions, water quality and sediment conditions, the status of environmental assets, and climate trends and extremes is expected to decline. The Mekong River Commission must find new news to promote joint understanding of the interconnectedness of the ecosystems and economies of the riparian states, leverage data and technology, highlight accomplishments, develop shared goals that are truly owned, and encourage material collaboration toward these. How might that be accomplished?

The Mekong River Commission should be turned into an authority to improve institutional and organizational arrangements for managing the biotic, land, and water resources of the Lower Mekong River Basin. Turning the Mekong River Commission into an authoritative body would require several steps:

- **Political Will.** The governments of Laos, Thailand, Cambodia, and Vietnam must have the political will to cede some of their national sovereignty over water resource management to a regional authority.
- **Institutional Capacity.** The Mekong River Commission would need to build its institutional capacity to handle the additional responsibilities that come with being an authoritative body. This relates to increased funding, staffing, and technical expertise.
- **Legal Framework.** Laos, Thailand, Cambodia, and Vietnam would need to agree on a new legal framework that grants the Mekong River Commission regulatory and enforcement powers. This would involve amending the existing 1995 Mekong Agreement.
- **Stakeholder Engagement.** The Mekong River Commission would need to engage with local communities, nongovernment organizations, and other stakeholders to ensure that its transition to an authoritative body is inclusive and encompasses the needs and concerns of all affected parties.
- **International Support.** Gaining support from international organizations and donor countries can provide the necessary resources and legitimacy to strengthen the Mekong River Commission's role.

Transforming the Mekong River Commission into an authoritative body is undoubtedly complex and challenging: even so, more likely than any other foreseeable means with time running out, it could enhance the sustainable management of the Mekong River Basin by ensuring consistent and enforceable policies across its lower region. If the elevation of the Mekong River Commission into an authority is not acceptable, what else—in the spirit of environmental peacebuilding—might the governments of Laos, Thailand, Cambodia, and Vietnam commit to in support of the vision and mission they subscribed to in 1995, and trumpeted again in their Vientiane Declaration as recently as 2023? "An ounce of practice is worth more than tons of preaching," Mahatma Gandhi reflected.

References

Britannica (2024) Mekong River. https://www.britannica.com/place/Mekong-River

Brown O, Nicolucci-Altman G (2022) White paper on the future of environmental peacebuilding. Geneva Peacebuilding Platform, International Union for Conservation of Nature, PeaceNexus Foundation, Environmental Law Institute, Environmental Peacebuilding Association.

Carius A (2008) Environmental cooperation as an instrument of crisis prevention and peacebuilding: conditions for success and constraints. Paper presented at the 2006 Berlin Conference on the Human Dimensions of Global Environmental Change "Resource Policies: effectiveness, Efficiency, and Equity", Berlin, November 17–18.

Carroll J (2018) Pulse of the forest: the state of the greater Mekong's forests and the everyday People working to protect them. World Wide Fund for Nature, July 19. https://www.worldwildlife.org/publications/pulse-of-the-forest-the-state-of-the-greater-mekong-s-forests-and-the-everyday-people-working-to-protect-them

Diamond J (2005) Collapse: how societies choose to fail or succeed. Viking Press

Dresse A, Fischhendler I, Nielsen J, Zikos D (2019) Environmental peacebuilding: towards a theoretical framework. Coop Confl 54(1):99–119

Eyler B, Kwan R (2024) All dams map of the Mekong basin. Stimson Center, May 7. https://www.stimson.org/2024/all-dams-map-of-the-mekong-basin/

Goo C (2021) China and the Mekong River disputes: can a new framework bring new compromises? China Focus, April 9. https://chinafocus.ucsd.edu/2021/04/09/china-and-the-mekong-river-disputes-can-a-new-framework-bring-new-compromises/

Haefner A (2020) 25 years of the Mekong River Commission: looking back, looking ahead. Australian Outlook, April 14. https://www.internationalaffairs.org.au/australianoutlook/25-years-of-the-mekong-river-commission-looking-back-looking-ahead/

Hutagalung S (2024) A secure future: environmental cooperation for peacebuilding in the 21st century. Eurasia Review, May 13. https://www.eurasiareview.com/13052024-a-secure-future-environmental-cooperation-for-peacebuilding-in-the-21st-century-oped/

Ide T, Bruch C, Carius A, Conca K, Dabelko G, Matthew R, Weinthal E (2021) The past and future(s) of environmental peacebuilding. Int Aff 97(1):1–16

Krampe F, Swain A (2021) Environmental peacebuilding. In: Richmond O, Visoka G (eds) The Oxford handbook of peacebuilding, statebuilding, and peace formation(online ed). Oxford Academic

Mekong River Commission (2017) 1995 Mekong agreement and procedures. https://www.mrcmekong.org/agreement

Mekong River Commission (2021) The integrated water resources management–based basin development strategy for the Lower Mekong Basin 2021–2030 and the MRC strategic plan 2021–2025. https://www.mrcmekong.org/strategy

Mekong River Commission (2024) Vision and mission. https://www.mrcmekong.org/vision-and-mission

Serrat O (2004) A note on Mekong River management. Unpublished manuscript, Asian Development Bank. https://www.researchgate.net/publication/291056409_A_Note_on_Mekong_River_Management

Serrat O (2017a) Reading the future. In: Knowledge solutions: tools, methods, and approaches to drive organizational performance. Springer, pp 11–14

Serrat O (2017b) Asking effective questions. In: Knowledge solutions: tools, methods, and approaches to drive organizational performance. Springer, pp 889–895

Serrat O (2023) Safeguarding Lake Chad. Unpublished manuscript, Georgetown University. https://www.researchgate.net/publication/376618564_Safeguarding_Lake_Chad

Shakespeare W (1993) Romeo and Juliet. Dover Publications. (Original work published 1597)

United Nations Environment Programme (n.d.) Vientiane declaration on innovation and cooperation for a water secure and sustainable Mekong. https://leap.unep.org/en/countries/kh/national-legislation/vientiane-declaration-innovation-and-cooperation-water-secure-and

United Nations General Assembly (2015) Transforming our world: the 2030 agenda for sustainable development. A/RES/70/1. https://sdgs.un.org/publications/transforming-our-world-2030-agenda-sustainable-development-17981

World Wide Fund for Nature (2024) Greater Mekong. https://www.worldwildlife.org/places/greater-mekong

Part VII
The Environment in Africa

Chapter 24
Safeguarding Lake Chad

Abstract Lake Chad is a biological hotspot and a source of food and water for millions of people in Central Africa. Lake Chad has shrunk by more than nine-tenths since the mid-1960s because of water diversion, population growth, and climate change. This précis considers the issues facing Lake Chad and tables a proposal to safeguard it.

24.1 The Setting

Located at the conjunction of Chad, Cameroon, Nigeria, and Niger in the Sahelian zone of Central Africa, Lake Chad supports unique ecosystems and biodiversity and provides food and water to approximately 30 million people. Alarmingly, Lake Chad lost most of its water over the past half-century and now spans about 5% of the area it covered in the 1960s (Hansen 2017; Serrat 2023). Lake Chad may vanish over the next 20 years. This proposal considers the issues facing Lake Chad, frames the question that must be addressed is Lake Chad is to be saved, and surmises what can be expected based on the assumed knowledge. Enmeshed in a worsening human and physical environment, Lake Chad is a symbolic case study in sustainable development: what with climate variability caused by climate change, a population explosion, and resource overuse, no single culprit explains the perfect storm of environmental and social pressures bearing on the lake. And so, if the shared blueprint of the Sustainable Development Goals is to find success, now and into the future, it truly behooves humankind to try to safeguard Lake Chad and its outstanding universal value (United Nations General Assembly 2015).

O. Serrat, *Anthropogenic Solutions for Climate Change*,
https://doi.org/10.1007/978-981-96-8654-4_24

24.2 Lake Chad: An Overview

24.2.1 The Partitioning of Africa

The Berlin Conference (1884–1885) regulated European colonization and trade in Africa and unleashed a veritable scramble for the continent's agricultural and mineral resources. From their capitals, European powers partitioned Africa with little knowledge of its ethnic composition, geography, or history. "We [the British and the French] have been engaged in drawing lines upon maps were no white man's foot ever trod: we have been giving away mountains and rivers and lakes to each other, only hindered by the small impediments that we never knew exactly where the mountains and rivers and lakes were," Anene (1970, p. 3) reported Lord Salisbury, the British Prime Minister (1885–1892), as saying. By 1900, Belgium, France, Germany, Great Britain, Italy, Portugal, and Spain had colonized most of Africa, with France and Great Britain at the forefront. It is a legacy of imperialism and the partitioning of Africa that, post-independence, four former colonies of France (Chad and Niger), Germany (Cameroon), and Great Britain (Nigeria) share access to Lake Chad and its fast-disappearing hydrospheric, biospheric, and lithospheric resources.

24.2.2 Hydrosphere

Of tectonic origin and with a catchment (or watershed) area of 2,500,000 square kilometers (965,000 square miles), Lake Chad is nestled in the low spot of the basin of a much larger ancient sea, viz., Mega Chad, that until around 5000 BCE drained into the Atlantic Ocean through the Benue River system in modern day Cameroon and Nigeria (Britannica 2023). Stretching approximately 400,000 square kilometers (155,000 square miles) across the Sahara, Mega Chad would have been the world's largest lake today, be it saline or freshwater, and 180 meters (600 feet) deep in places (Britannica 2023; Express Digest 2020). Comparing, the largest saline lake is now the Caspian Sea (386,400 square kilometers or 149,200 square miles) and the largest freshwater lake is now Lake Superior (82,103 square kilometers or 31,700 square miles); currently, the largest lake in Africa is Lake Victoria (59,947 square kilometers or 23,146 square miles) (World Atlas 2023).

Today, Lake Chad is a variable body of freshwater with a surface area that peaks in October–November and shrinks by more than half in April–May (Lakepedia n.d.). Lake Chad has no important outflows: its volume is conditioned by the discharge of its catchment, balanced by evaporation, transpiration, and seepage (Britannica 2023). Lake Chad benefits mainly from the Chari (Shari) River via the Logone tributary—sourced from the western area of the Central African Republic, northern Cameroon, and southern Chad—which accounts for about 90–95% of inflow; the rest is contributed by the Komadugu-Yobe River and smaller tributaries such as the Ngadda River and Yedseram River, sourced from Nigeria (Britannica 2023). Lake Chad is shallow and averages 1.5 m (5 feet) in depth, with maximum depths of 4–8 meters (13–26 feet) in the north and 2–4 meters (6.5–13 feet) in the south (Lakepedia n.d.).

Lake Chad lies in the central semiarid tropical (Sahel) zone of Africa. In the short rainy season from July to September, highly variable average annual precipitation in the Lake Chad area amounts to about 330 mm, with higher precipitation in the south than in the north. The annual maximum temperatures are as high as 35–40°C (95°F–104°F), particularly in the north, and droughts are increasingly common. As of 10 years ago, according to the U.S. Geological Survey (2012), temperatures had already increased by 0.8°C (33°F) over those in 1975, consequently exacerbating the effects of droughts. Currently, the length of Lake Chad's shoreline extends over 500–800 kilometers (310–500 square miles) (Lakepedia n.d.).

Until the 1960s, Lake Chad was the fourth largest lake in Africa and the sixth largest lake in the world: however, its surface area has since dwindled from 26,000 square kilometers (10,000 square miles) in 1963 to less than 1500 square kilometers (580 square miles), or about 5% of its original area (Lakepedia n.d.). In the mid-1970s, Lake Chad transitioned from one body of open water to a different configuration dominated by marshlands spreading over two lobes of much smaller sizes. The northern and southern lobes of Lake Chad have not reconnected since 1973 (Hansen 2017).

24.2.3 Biosphere

The soils around Lake Chad once supported woodland but human activities and reduced moisture from water diversion and climate change have reduced the diversity of plant life; what remains includes acacias, baobab, desert date, Indian jujube, African myrrh, and palms (Britannica 2023). The inundated lands that surround Laker Chad are more vegetated: annual grasses such as ambatch, papyrus, reeds, and water lilies dominate (Britannica 2023). Habitat loss, hunting, and competition from livestock have depleted wildlife: rhinoceroses have been reduced but hippopotamuses and crocodiles still inhabit Lake Chad's waters (Britannica 2023). Large carnivores such as leopards and lions have been eliminated; rodents have benefited from these changes (Britannica 2023). There are many water and shore birds (e.g., fulvous tree ducks, garganeys, Egyptian geese, glossy ibises, pink-backed pelicans, shovelers, African spoonbills, marabou storks) as well as terrestrial birds (e.g., Nubian bustards, ground hornbills, ostriches, secretary birds) (Britannica 2023). Amphibians and reptiles include Nile crocodiles, rock pythons, and spitting cobras. Lake Chad remains an important fishery, with about 40 species of commercial importance; as with plant life, however, the trend is toward decreased diversity and lower levels of biological productivity (Britannica 2023).

24.2.4 Lithosphere

The geological basin of Lake Chad harks to the Phanerozoic Era (from 538 million years ago to the present): it is a sedimentary basin formed during the plate divergence that opened the South Atlantic Ocean (Wikipedia n.d.). The Lake Chad Basin contain concentrations of fossils and oil exceeding 1.5 billion barrels of proven reserves was discovered north of Lake Chad (Britannica 2023). Exploitation by Western companies such as Exxon-Mobil and Royal Dutch Shell began in 2003 and reached about 140,000 barrels per day in 2020 through a pipeline from Chad to Cameroon (International Trade Administration 2020; Wikipedia n.d.). Oil is Chad's primary source of public revenue (International Trade Administration 2020). Natron, a complex sodium carbonate used as salt and in the preparation of soap and medicine, is another mineral resource of the Lake Chad area (Britannica 2023). Gold is found in various parts of Chad and is mined in the southwestern part of the country (Britannica 2023). Other mineral deposits are bauxite, titanium, and uranium (Britannica 2023).

24.2.5 People and Economy

People have lived in the Lake Chad Basin for at least 8000 years (Wikipedia n.d.). From about 7 million in the 1960s, the population of the Lake Chad Basin has experienced rapid growth in the last decades (2.5–3.0%) and is currently estimated to be over 30 million people, most of whom eke a living on the south bank of Lake Chad where population density exceeds 100 people per square kilometer (259 people per square mile) (Wikipedia n.d.). There are more than 70 ethnic groups around Lake Chad (Wikipedia n.d.). All rely on Lake Chad for food and water security and conflicts are escalating, with the four countries along the lake facing extreme poverty (Wikipedia n.d.). According to the United Nations Development Programme (UNDP 2022), Chad was ranked 190th out of 191 countries in the United Nations' 2021–2022 Human Development Index.

The population in the Lake Chad area depends for 80–90% of its livelihoods on subsistence (and some commercial) agriculture, fisheries, and animal husbandry, usually in combination (Britannica 2023). Crops consist of beans, corn, millet, sorghum, and vegetables (e.g., cowpea, gourd); some wheat is grown in polders based on traditional agricultural practices. Crop yields are very low and stagnant. Fishing is the most visible activity on Lake Chad and provides about 45% of household income but statistics are uncertain (Britannica 2023). In the 1960s, Lake Chad was home to 135 species of fish and fishermen landed 200,000 metric tons every year according to the United Nations Environment Programme (UNEP 2018). In the 1980s, total production of fish was estimated at 60,000–120,000 tons per year; more recent annual harvests are reckoned at 50,000–60,000 tons, with a corresponding decrease in biodiversity (Sherwood 2022). It is estimated that 50% of all fish landed is exported: smoked or sundried fish called *salanga* and larger smoked fish called *banda* are marketed in Cameroonian and Nigerian markets, respectively (Britannica 2023). Cattle are the most important livestock raised: a notable breed used for meat, milk, and work is the Kuri, a unique humpless longhorn found only in the Lake Chad area (UNEP 2018). Asses, camels, goats, horses, poultry, sheep, and several varieties of zebu are also kept (Britannica 2023).

24.2.6 Transportation

Watercraft made of papyrus-reed or sewn planks are used for fishing but Lake Chad is little used for commercial navigation (Britannica 2023). In the 1950s, there was intermittent barge traffic between Bol (to the north of the southern lobe of Lake Chad) and N'Djamena (to the south), presumably also using the Chari River (Britannica 2023). As of 2012, the Chari River and the Logone River were only navigable in the wet season (Wikipedia n.d.).

24.2.7 The Lake Chad Basin

The hydrologically closed catchment of Lake Chad covers 2,500,000 square kilometers (965,000 square miles), equivalent to about 8% of the African continent, and straddles eight countries: Algeria, Libya, Niger, Chad, Sudan, Nigeria, Cameroon, and the Central African Republic (Wikipedia n.d.). The Lake Chad Basin is surrounded by mountains: it has no outlet to the sea and encompasses vast areas of semi-arid desert and savanna. Table 24.1 presents key geographical data on the Lake Chad Basin.

Table 24.1 The Lake Chad Basin

	Total area (km^2)	Area within basin (km^2)	% of total area of basin	% of country in basin
Chad	1,284,000	1,046,196	43.8	81.5
Niger	1,267,000	691,473	28.9	54.6
Central African Republic	622,984	219,410	9.2	35.2
Nigeria	923,768	179,282	7.5	19.4
Sudan	475,440	50,775	2.1	10.7
Algeria	1,861,484	101,048	4.2	5.4
Cameroon	2,381,740	93,451	3.9	3.9
Libya	1,759,540	7065	0.3	0.4
Total	10,575,956	2,388,700	100.0	

Note: Compiled from FAO (1997), Statistics Times (2020)

24.3 Lake Chad: Key Challenges

24.3.1 Developmental Pressures

The Lake Chad Basin Commission (Africa Renewal 2019) and the United Nations Environment Programme (UNEP 2018) have charged that inefficient damming and unsustainable irrigation in the countries bordering the lake are partly responsible for its shrinkage. "We are losing a lot of plant life and animals, including fish species. Farmers, herders, and fishermen have lost their livelihoods. It's a real crisis for us," said the Executive Secretary of the Lake Chad Basin Commission (Africa Renewal 2019, para. 1). The World Wildlife Fund has specifically pointed to the diversion of water from the Chari River to irrigation projects and dams along the Jama'are and Hadejia Rivers, which later form the Yobe River in northeastern Nigeria (Africa Renewal 2012). Significantly, Coe and Foley (2001) determined that irrigation had a modest impact on Lake Chad over the period 1953–1979 but that it had increased fourfold over the period 1983–1994. Pointedly, using an integrated biosphere model and a hydrological routing algorithm in conjunction with long time-series climate data, Coe and Foley (2001) attributed about 50% of the decrease in Lake Chad's size since the 1960s to human water use; Coe and Foley (2001) ascribed the rest to shifting climate patterns that may also have been affected by loss of vegetation from overgrazing hence a drier climate. As Lake Chad continues to dry up, farmers and cattle herders move to greener areas where they compete for land resources with other communities; some migrate to cities (e.g., Abuja, Kano, Lagos) for menial jobs or roam the streets as beggars (Africa Renewal 2012).

24.3.2 Climate Change

A population explosion and resulting resource overuse is not breaking news but the receding waters of Lake Chad also owe to a reduction in local precipitation, induced by climate change. According to the Intergovernmental Panel on Climate Change, as reported by the Deutsche Gesellschaft für Internationale Zusammenarbeit (GIZ 2015), climate change is set to have particularly serious consequences in the Sahel. Climate change affects all manner of natural capital (e.g., biodiversity, food, forest, land, water). Slowly, the Lake Chad area is turning into a desert landscape: as the waters recede, silt and sediments on the lakebed dry in the sun and are swept up by wind gusts over hundreds, sometimes even thousands, of kilometers. In an area where most of the population lives from subsistence agriculture, fisheries, and animal husbandry, the predicted 2°C increase by 2050 could lead to a 15–20% reduction in food production. According to the International Committee of the Red Cross, as cited by the United Nations Security Council (2020), the communities of the Lake Chad area "now walk on a tightrope of survival."

24.3.3 Conflict

The shrinking of Lake Chad, the ever-greater variability in local precipitation due to climate change, and rising temperatures have eroded social cohesion and trust and encouraged different types of self-help survival measures. Arms trafficking and smuggling, facilitated by the porous borders between Cameroon, Chad, Niger, and Nigeria, is on the rise; youths are recruited by bandits; and increasingly violent clashes take place between farmers and herders in consequence of cross-border transhumance. The uprising of Boko Haram, a militant Islamist group operating from the north-eastern Borno State of Nigeria since 2009, has added further pressures on Lake Chad (Al Jazeera 2021). Conflict affects agricultural, fishing, and pastoral livelihoods; markets and trade; and other income generating activities as people are unable to access land or Lake Chad or lose farming or fishing equipment, crops are destroyed or harvested by armed opposition groups, markets are closed, and people are relocated by security forces (Crop Monitor 2020). Over 2.3 million people across the region are displaced; over 5 million are struggling to access enough food to survive; and half a million children are suffering from severe acute malnutrition, according to the UN High Commissioner for Refugees as reported by Al Jazeera (2021), because of a conflict that shows little sign of ending despite the efforts of a joint multinational task force from Benin, Cameroon, Chad, Niger, and Nigeria.

24.3.4 Governance

Sustainable development hinges on good governance. On May 22, 1964, the leaders of Chad, Cameroon, Niger, and Nigeria—the four countries bordering Lake Chad—established the all-natural Lake Chad Basin Commission to help manage the waters of Lake Chad (Lake Chad Basin Commission 2023). The Central African Republic joined the Lake Chad Basin Commission in 1994 and Libya was admitted in 2008; Egypt, the Democratic Republic of Congo, the Republic of Congo, and Sudan hold observer status (Lake Chad Basin Commission 2023). The Lake Chad Basin Commission is a member of both the African Network of Basin Organizations and the International Network of Basin Organizations; it is funded by contributions from its six member states, with headquarters in N'Djamena, Chad (Lake Chad Basin Commission 2023). The mandate of the Lake Chad Basin Commission is to:

- Manage of the waters of Lake Chad and the other transboundary water resources in a sustainable and equitable way.
- Conserve the ecosystems of the conventional basin.
- Promote regional integration and safeguard peace and security in the conventional basin (Lake Chad Basin Commission 2023).

To fulfill its mandate, the Lake Chad Basin Commission is to (a) engage in water resources management; (b) take climate action for ecosystem and biodiversity protection; and (c) promote regional integration, peace, and security (Lake Chad Basin Commission 2023). But, good governance by the Lake Chad Basin Commission faces seemingly insurmountable challenges:

- **Development Policies.** With only about 5% of the surface area that Lake Chad had in the 1960s remaining, development policies that focus on short-term solutions, such as additional water diversion for irrigation, are a grave concern. Large-scale structural solutions such as Transaqua may be last-ditch efforts.
- **Population Growth and Displacement.** The Lake Chad basin is home to about 30 million people, many of them among the poorest in the world. The combined effects of resource overuse, climate change, and desertification have displaced populations and worsened food insecurity. Lake Chad region's population, which has doubled over the past few decades, could double again over the next 20 years. Taub (2017) labelled Lake Chad the world's most complex humanitarian disaster.
- **International Cooperation.** Managing the shared water resources of Lake Chad across different countries is not easy, as other basin organizations such as the Mekong River Commission—which works with the governments of Cambodia, the Lao People's Democratic Republic, Thailand, and Viet Nam—have found (Serrat 2004a).
- **Security.** Beginning in the 1980s, armed criminal groups and Islamic extremists have taken refuge around Lake Chad, increasing stress on natural resources and aggravating population displacement.
- **Financial Challenges.** The Lake Chad Basin Commission's annual budget is funded by contributions from the member states, but not Libya, based on an agreed-upon formula: Nigeria provides 52% of the annual budget, followed by Cameroon (26%), Chad (11%), Niger (7%), and the Central African Republic (4%). However, the annual budget of the Lake Chad Basin Commission amounts to a mere $1 million: what can the Lake Chad Basin Commission hope to accomplish with such meager resources?

Left unaddressed, challenges such as those mentioned above will compound over time, making the task of preserving Lake Chad evermore complex. The vision of the Lake Chad Basin Commission, to be realized by 2025, is one where (a) Lake Chad is maintained at sustainable levels; (b) regional and national authorities accept responsibilities for freshwater, ecosystem, and biodiversity conservation and for judicious integrated river basin management to achieve sustainable development; and (c) every member state has equitable access to safe and adequate water resources to meet its needs and rights and maintain its freshwater, ecosystem, and biodiversity resources (Lake Chad Basin Commission 2023). And yet, most probably because its annual budget is so small, the specific missions with which the Lake Chad Basin Commission is tasked to achieve the said vision entail little other than (a) gathering, examining. and disseminating information about the projects drawn up by its member states and recommending a common work plan and joint research programs in the Lake Chad basin; (b) maintaining the link between the High Contracting Parties, ostensibly to ensure effective use of water resources; (c) monitoring the implementation of studies and work in Lake Chad basin and informing member states in this regard; (d) drawing up a common set of rules to regulate navigation; (e) establishing regulations that cover Lake Chad Basin Commission staffing, and ensuring that these are applied; and (f) reviewing complaints, encouraging the settlement of disputes, and reinforcing regional cooperation (Lake Chad Basin Commission 2023). The Lake Chad Basin Commission has no political traction: this explains why it is poorly funded and unable to deliver on its mandate. In 2018, a representative of the Lake Chad Basin Commission and numerous other speakers reported to the United Nations Security Council (2018) that better governance of the poorly funded and badly managed Lake Chad Basin Commission was key to resolving conflict and suffering across the Lake Chad area. Without a doubt, 60 years after it was established, there is no chance that the Lake Chad Basin Commission's vision for Lake Chad will be realized by 2025.

24.4 Safeguarding Lake Chad: A Proposal

24.4.1 Review of Past Approaches

The Lake Chad area holds unparalleled ecological significance and is home to 30 million people. Despite that, the Lake Chad area has attracted little investment from its riparian states or from technical and financial partners such as the African Development Bank (AfDB), the Agence Française de Développement (AFD), the European Union, GIZ, the Global Environment Facility, and the World Bank. UNDP, UNEP, the United Nations Educational, Scientific and Cultural Organization (UNESCO), and other bodies of the United Nations have extended some technical assistance. A desk review of past approaches reveals siloed projects, each reflecting the orientations of donors. To illustrate, the Lake Chad Basin Commission developed a $904 million five-year investment program for 2013–2017 to reduce poverty and improve the living conditions of people. The plan was presented at a donors' roundtable on April 4–5, 2014 in Bologna, Italy but most of its activities remained unfunded. Another five-year investment program to which several bilateral and multilateral donors subscribed aimed to "fast-track" $300 million by 2020, a modest amount if disbursed across the six member states of the Lake Chad Basin Commission (World Bank 2020).

24.4.2 Transaqua

First proposed in the 1970s, Transaqua is a large-scale water infrastructure project that calls for the construction of a 2400-kilometer (1490 miles) canal to replenish Lake Chad with water transferred from the Congo River Basin (Nagabhatla and

Sayan 2020). Transaqua would be one of the biggest water projects in the world. The construction works for Transaqua would take place in the Central African Republic and, chiefly, the Democratic Republic of Congo, not in the riparian states of the Lake Chad Basin. At the International Conference on Lake Chad held in Abuja, Nigeria from February 26 to March 1, 2018, the Lake Chad Basin Commission endorsed Transaqua as a preferred option for water transfer to restore Lake Chad after state-owned PowerChina and Bonifica, a consulting and engineering company, offered to support the investment (Nagabhatla and Sayan 2020). Key players including the World Bank, the Commission Internationale du Bassin Congo-Oubangui-Sangha (the river basin organization managing the Congo River), nongovernment organizations such as International Rivers, and riparian states of the Congo River have not disclosed their positions (Nagabhatla and Sayan 2020). The Democratic Republic of Congo, which would host most of the construction works to allocate water from its territories, has predictably raised concerns.

24.4.3 The Rationale for Action

The hypothesis, viz., the supposition made based on what evidence was collected in the précis as a starting point for further investigation, is that Lake Chad could disappear within the next 20 years. If so, the question is: Can Lake Chad be saved? Every day, in a veritable regional crisis, tens of millions of people around Lake Chad struggle to survive in the face of climate change, corruption, depletion of natural resources, food shortages, poor access to essential public services, sectarian mistrust, and water insecurity, all of which have fostered terrorism and the growth of Boko Haram in the region (Wikipedia n.d.). The drying of Lake Chad is affecting real lives. Local and regional communities as well as their leaders should care about safeguarding Lake Chad; but so should, just as intently, decision makers, the donor community, international organizations, nongovernment organizations, policy makers, and society in Africa and around the world.

Lake Chad is a case study in sustainable development, a "Gordian knot of problems" according to the United Nations (2023). Remarkably, however, the convergence of the challenges affecting Lake Chad could yield synergistic improvements across the entirety of the Sustainable Development Goals that the General Assembly of the United Nations endorsed on September 15, 2015 to create a better world by 2030 in the five critical areas of people, planet, prosperity, peace, and partnership (United Nations General Assembly 2015). With no exception, the Sustainable Development Goals can all be made to work for Lake Chad:

- SDG 1: No Poverty
- SDG 2: Zero Hunger
- SDG 3: Good Health and Well-Being
- SDG 4: Quality Education
- SDG 5: Gender Equality

- SDG 6: Clean Water and Sanitation
- SDG 7: Affordable and Clean Energy
- SDG 8: Decent Work and Economic Growth
- SDG 9: Industry, Innovation, and Infrastructure
- SDG 10: Reduced Inequalities
- SDG 11: Sustainable Cities and Communities
- SDG 12: Responsible Consumption and Production
- SDG 13: Climate Action
- SDG 14: Life Below Water
- SDG 15: Life on Land
- SDG 16: Peace, Justice, and Strong Institutions
- SDG 17: Partnerships for the Goals

24.4.4 A Proposal

Governance has to do with the laws, norms, policies, practices, and rules that direct human behavior. Necessarily, therefore, governance must address such matters as who makes decisions, how scientific information for decision-making is generated, how decisions are made and carried out, and how stakeholders can participate in the process. A commission is appointed, usually on a temporary basis, to carry out specific tasks or investigations.

Quite differently, an authority is an organization or body that, buttressed by enabling acts granting it legitimacy, has the power to make decisions and enforce rules in a specific field or area. If Lake Chad is to be saved the Lake Chad Basin Commission should be turned into an authority, charged with managing the lake in an integrated and sustainable manner from a regional perspective (Serrat 2003, 2004b). The Tennessee Valley Authority, established in 1933, is one example of an agency that fused a broad range of specific powers in its region. The objective of the Lake Chad Basin Authority would be to establish and act on improved institutional and organizational arrangements for managing the biotic, land, and water resources of the Lake Chad Basin. To mobilize and galvanize support from a wide range of partners, the six member states of the Lake Chad Basin Commission would need to promptly finalize the design of the Lake Chad Basin Authority and the design of sub-basin institutional structures, especially regarding power, duties, and necessary staff.

24.4.5 An Approach: Future Search Conferencing

Toward the establishment of the Lake Chad Basin Authority, the Lake Chad Basin Commission should hold a Future Search Conference on Safeguarding Lake Chad to formulate a clear and powerful image of the future and enlist commitment. Future

Search conferencing is a system-wide, task-focused, strategic planning tool that enables potentially conflicting groups to find ground for common, constructive action (Weisbord 1992; Serrat 2017). Future Search conferences are held over three days to help groups of 60–100 people identify a shared vision of the future. Specifically, in four or five sessions each lasting half a day, participants (a) focus on the past (i.e., highlights and milestones), (b) focus on the present (i.e., external trends), (c) focus on the present (i.e., responses to trends, owning actions), (d) focus on the future (i.e., ideal scenarios), (e) discover common ground, (f) confirm common ground, and (g) plan action (Weisbord 1992; Serrat 2017). A key output of the Future Search Conference on Safeguarding Lake Chad would relate to the elaboration of the institutional framework for the Lake Chad Basin Authority including (a) a constitutional mandate for the Lake Chad Basin Authority, (b) the areas of responsibility of the Lake Chad Basin Authority (including a robust and effective regulatory framework and operating protocol to underpin basin management), (c) the structure of the Lake Chad Basin Authority (including the means by which it would interact stakeholders), (d) capacity building requirements, and (e) the technical and financial support needed.

The local, national, regional, and global dimensions of the crisis impacting Lake Chad require that the Future Search conference give prominence to representatives of the African Union, the Economic Community of West African States, local and regional communities, policy makers, decision makers, the donor community, the International Network of Basin Organizations, international organizations, and non-government organizations. The African Network of Basin Organizations, a key actor for cooperation within the African Union that aims to strengthen the capacity of African basin organizations to effectively plan and implement joint multisectoral basin-wide programs for sustainable regional economic development, would chair the Future Search conference. The Lake Chad Basin Commission, which is a member of both the African Network of Basin Organizations and the International Network of Basin Organizations, would serve as facilitator and logistics manager.

24.5 Summary, Conclusion, and Recommendation

24.5.1 Summary

Lake Chad is a biological hotspot and a source of food and water for 30 million people in Central Africa; however, it has shrunk by more than nine-tenths since the mid-1960s because of water diversion, population growth, and climate change. Once the largest lake in the world, Lake Chad may vanish over the next 20 years. Relentlessly, the shrinking of Lake Chad has compromised the economic, ecosystem, peace and security, as well as regional and economic benefits that humans used to derive from a once-great lake. Lake Chad is an ecological catastrophe: its final demise would compound what is already a humanitarian disaster.

24.5.2 *Conclusion*

Desperate times call for desperate measures. The Lake Chad Basin Commission has not been able to deliver on its mandate. The six member states of the Lake Chad Basin Commission should design their chief governance instrument for success, not failure: they should invite national, regional, and global stakeholders to deploy relevant, coherent, effective, efficient, impactful, and sustainable partnerships that place emphasis on involving stakeholders, strategic long-term planning, and basin action plans, not forgetting communication (i.e., raising awareness, education) (Global Water Partnership & International Network of Basin Organizations 2009).

24.5.3 *Recommendation*

The Lake Chad Basin Commission should be turned into an authority to establish improved institutional and organizational arrangements for managing the biotic, land, and water resources of the Lake Chad Basin. Toward that, a Future Search conference should be held to develop a series of options for the future and agree on a plan of action in the short, medium, and long term; the plan of action should detail long-drawn-out capacity building of the proposed Lake Chad Basin Authority to optimize efficiency and maximize effectiveness. Of course, this is not to underestimate the formidable difficulties that preserving Lake Chad presents, not least of which with respect to the mustering of political will, the forging of bilateral and multilateral agreements, the commitment of long-term finance, and the setting up of supportive national and national institutions. Addressing the foregoing goes without saying and related measures would constitute integral part of the plan of action because the time for siloed and woefully inadequate project-based approaches is over and farfetched structural solutions à la Transaqua will not do. "Change will not come if we wait for some other person or if we wait for some other time. We are the ones we've been waiting for. We are the change that we seek," Obama (2008) made clear. Sixty years after the Lake Chad Basin Commission was established, what with only about 5% of Lake Chad remaining, yesterday is long gone and tomorrow is today.

References

Africa Renewal (2012) Africa's vanishing Lake Chad, April 15. https://www.un.org/africarenewal/magazine/april-2012/africa%E2%80%99s-vanishing-lake-chad

Africa Renewal (2019) Address development issues in the Lake Chad basin, December 24. https://www.un.org/africarenewal/magazine/december-2019-march-2020/address-development-issues-lake-chad-basin

Al Jazeera (2021) Northeast Nigeria conflict killed more than 300,000 children: UN, June 24. https://www.aljazeera.com/news/2021/6/24/northeast-nigeria-conflict-killed-more-than-300000-children-un
Anene J (1970) The international boundaries of Nigeria 1885–1960: the framework of an emergent African nation. Humanities Press
Britannica (2023) Lake Chad. https://www.britannica.com/place/Lake-Chad
Coe M, Foley J (2001) Human and natural impacts on the water resources of the Lake Chad basin. J Geophys Res 106(D4):3349–3356
Crop Monitor (2020) Lake Chad basin: conflict and food insecurity. University of Maryland, August 13. https://cropmonitor.org/documents/SPECIAL/reports/Special_Report_20200801_Lake_Chad_Basin.pdf
Express Digest (2020) NASA image reveals remains of an ancient lake that stretched across the Sahara 7,000 years ago, February 20. https://expressdigest.com/nasa-image-reveals-remains-of-an-ancient-lake-that-stretched-across-the-sahara-7000-years-ago/
FAO (1997) Irrigation potential in Africa: a basin approach. https://www.fao.org/3/w4347e/w4347e00.htm
GIZ (2015) Adapting to climate change in the Lake Chad basin. https://www.giz.de/en/worldwide/24845.html
Global Water Partnership & International Network of Basin Organizations (2009) A handbook for integrated water resources management in basins. https://www.gwp.org/globalassets/global/toolbox/references/a-handbook-for-integrated-water-resources-management-in-basins-inbo-gwp-2009-english.pdf
Hansen K (2017) The rise and fall of Africa's great lake. NASA Earth Observatory, November 9. https://earthobservatory.nasa.gov/features/LakeChad
International Trade Administration (2020) Chad: country commercial guide. https://www.trade.gov/country-commercial-guides/chad-oil-and-gas
Lake Chad Basin Commission (2023) Lake Chad basin commission. https://cblt.org/
Lakepedia (n.d.) Lake Chad: the shrinking giant. https://www.lakepedia.com/lake/chad.html
Nagabhatla N, Sayan R (2020) Transaqua, Lake Chad, and the Congo basin: a call for cautious action. ReliefWeb, November . https://reliefweb.int/report/democratic-republic-congo/transaqua-lake-chad-and-congo-basin-call-cautious-action
Obama B (2008) Super Tuesday [Speech Transcript]. Best Speeches of Barack Obama Through His 2009 Inauguration, February 5. http://obamaspeeches.com/E02-Barack-Obama-Super-Tuesday-Chicago-IL-February-5-2008.htm
Serrat O (2003) Establishment of the Tonle Sap basin management organization. Asian Development Bank. https://www.researchgate.net/publication/346952758_Establishment_of_the_Tonle_Sap_Basin_Management_Organization
Serrat O (2004a) A note on Mekong River management. Unpublished. https://www.researchgate.net/publication/291056409_A_Note_on_Mekong_River_Management
Serrat O (2004b) Establishment of the Tonle Sap basin management organization II. Asian Development Bank. https://www.researchgate.net/publication/346954110_Establishment_of_the_Tonle_Sap_Basin_Management_Organization_II
Serrat O (2017) Future search conferencing. In: Knowledge solutions: tools, methods, and approaches to drive organizational performance. Springer, pp 229–236
Serrat O (2023) Safeguarding Lake Chad [PowerPoint slides]. Georgetown University. https://www.researchgate.net/publication/376612603_Safeguarding_Lake_Chad
Sherwood M (2022) Environmental and sociopolitical aspects of Lake Chad, August 17. https://storymaps.arcgis.com/stories/3ccb2a0a3e9249f9be4cd92bf139c439
Statistics Times (2020) List of African countries by area. https://statisticstimes.com/geography/african-countries-by-area.php
Taub B (2017) Lake Chad: the world's most complex humanitarian disaster. The New Yorker, November 27. https://www.newyorker.com/magazine/2017/12/04/lake-chad-the-worlds-most-complex-humanitarian-disaster

U.S. Geological Survey (2012) A climate trend analysis of Chad, June 21. https://www.usgs.gov/publications/a-climate-trend-analysis-chad
UNDP (2022) Uncertain times, unsettled lives: shaping our future in a transforming world. Human Development Report 2021/2022.
UNEP (2018) The tale of a disappearing lake, February 28. https://www.unep.org/news-and-stories/story/tale-disappearing-lake
United Nations (2023) UN convenes Lake Chad countries, amid growing regional crisis. UN News, January 23. https://news.un.org/en/story/2023/01/1132742
United Nations General Assembly (2015) Transforming our world: the 2030 agenda for sustainable development. A/RES/70/1.
United Nations Security Council (2018) Better governance of underfunded, poorly managed Lake Chad basin key to resolving conflict, suffering across region, speakers tell Security Council, March 22. https://press.un.org/en/2018/sc13259.doc.htm
United Nations Security Council (2020) Sahel, Lake Chad region walking 'tightrope of survival' as result of environmental degradation, humanitarian organization president tells security council, September 17. https://press.un.org/en/2020/sc14307.doc.htm
Weisbord M (1992) Discovering common ground: how future search conferences bring people together to achieve breakthrough innovation, empowerment, shared vision, and collaborative action. Berrett-Koehler Publishers
Wikipedia (n.d.) Chad basin. https://en.wikipedia.org/wiki/Chad_Basin
World Atlas (2023) The largest freshwater lakes in the world. https://www.worldatlas.com/lakes/the-largest-freshwater-lakes-in-the-world.html
World Bank (2020) The next generation Africa climate business plan: ramping up development-centered climate action. https://openknowledge.worldbank.org/entities/publication/e44b41cc-9835-5acb-bce5-d5718bccb7bb

Chapter 25
The Demise of Hill Torrent Irrigation in Southern Punjab, Pakistan

Abstract With special attention to hill torrent management in southern Punjab, Pakistan, this précis recounts how the centralized irrigation schemes that Pakistan inherited from colonialism disturbed a hitherto balanced human and physical geography to ultimately leave the country's poor more vulnerable to critical trends, seasonalities, and shocks. The précis draws a short parallel with the colonial experience of southern Africa, where the production of cash crops for export also ruptured ecosystems, not forgetting the cultures and livelihoods of indigenous populations. The consequences that colonial rule had on indigenous production systems are being felt to this day in the former colonies and across the globe.

Covering 881,913 square kilometers (340,509 square miles), the Islamic Republic of Pakistan is the 33rd largest country in the world (viz., twice the size of California). With about 243 million people (2023), however, Pakistan is home to the world's fifth biggest population (and its second biggest Muslim population). Pakistan's multiethnic and multicultural society speaks Punjabi (about 39% of the population), Pashto (about 18%), Sindhi (about 15%), Saraiki (about 12%), Urdu (about 7%), or Balochi (about 3%). (Many other languages are spoken by smaller populations. English is used in official business and in government.) The administrative units of Pakistan comprise four provinces (viz., Punjab, Sindh, Khyber Pakhtunkhwa, and Baluchistan), one federal territory, and two disputed territories (Jammu and Kashmir). Vitally, the Indus River (1609 kilometers or 1000 miles) and its tributaries flow through Pakistan from Kashmir to the Arabian Sea, providing key water resources for the breadbasket of Punjab, which accounts for most of the nation's agricultural production, as well as Sindh.

More than half of Pakistan's population lives in Punjab (205,345 square kilometers or 79,284 square miles). Contributing about 60% of total agricultural production, Punjab is the largest producer of cotton, maize, rice, sugarcane, and wheat in the country from only about 26% of Pakistan's total land mass. (Nationwide, these four crops account for more than 75% of the value of Pakistan's total crop output.) Irrigation is the lifeline of Punjab's agriculture, with the Indus River (and its

---

The original version of the chapter has been revised. A correction to this chapter can be found at https://doi.org/10.1007/978-981-96-8654-4_27

O. Serrat, *Anthropogenic Solutions for Climate Change*,
https://doi.org/10.1007/978-981-96-8654-4_25

tributaries) serving as the primary sources of water through a network of dams, canals, and reservoirs that Great Britain introduced in colonial times. Where surface water is short, groundwater is tapped using tube wells. Zahid (2023) noted that the irrigation system has for long faced recurring challenges:

- **Climate Change.** Rising temperatures and changing precipitation strain the irrigation system, affecting the availability of water (quantity and timing) and leading to crop failures and food insecurity.
- **Water Scarcity.** The Indus River (and its tributaries) depend on snowmelt and monsoon rains, which are unpredictable. Both droughts and floods are common, with severe impacts on the irrigation system. Even so, over-extraction of water for irrigation frequently depletes water resources.
- **Deterioration of Infrastructure.** The dams, canals, and reservoirs are in poor condition, leading to water losses through seepage and breaches and then reduced crop yields. Certain areas are over-irrigated. Tubewell irrigation has led to salinization of land and water resources. Lack of financing limits the ability to maintain or repair infrastructure.

The hill torrents of southern Punjab, specifically Dera Ghazi (D.G.) Khan and Rajanpur Districts, are a source of insights into a deep-rooted problem affecting agriculture in Punjab, namely, colonial legacy fueled by enduring technocentrism (Merrey 1986). The Sulaiman Range is a mountain mass in central Pakistan that extends in a roughly north-south direction for about 450 kilometers (280 miles) and separates Punjab from Baluchistan. Its heights decrease toward the south with summits averaging 1800–2100 meters (6000–7000 feet). In Punjab, the piedmont of the Sulaiman Range is called *Pachad*, the Saraiki word for "west": it extends over a 30-kilometer wide (19 miles) and 40-kilometer long (25 miles) area (Shafiq 2012). In the monsoon months (July to September), flash floods known as *Rodh Kohi* (*Rodh* meaning "stream" and *Kohi* meaning "hill") flow into the barren alluvial fan, where spate irrigation (viz., torrent agriculture) is practiced, before draining into the Indus River (Kamran and Shamshad 2014; Shafiq 2012). *Rodh Kohi* systems go back at least as early as 330 BCE and sustained some of the early Indus Valley civilizations with indigenously developed, managed, maintained techniques. The *Pachad* has no canals and the entire belt depends for agriculture on 13 major hill torrents (and 200 smaller ones) (Qureshi et al. 2016). Hill torrent irrigation involves the construction of earthen diversion structures, to convey flood water across large distances, and earthen embankments (bunds) of up to 1.8 meters (6 feet) in height to trap and store it (Qureshi et al. 2016). Historically, the system rested on collective ownership by villages and community participation in the construction and maintenance of the diversion structures and embankments (Kamran and Shamshad 2014). Crops would be sown upon the drying of water and thrive on the moisture stored in the soil (Shafiq 2012). There was no further irrigation, save the occasional rain if any occurred. The major crops were millet, typically covering about 40% of the cropped area, followed by wheat (about 23%), gram (about 22%), and other crops (about 15%) (Shafiq 2012). The priority of the *Pachad*'s transhumant pastoralist tribes were fodder for their cattle, followed by wheat for their own subsistence, and finally a cash crop; livestock, especially goat and sheep, were fed millet and other foliage (Shafiq 2012). The tribes would migrate seasonally between the hills, where they engaged in pastoralism, and the alluvial fan that the hill torrents irrigated (Raza 2023). Seasonal migration was in harmony with a local ecology shaped by the periodicity of flooding (Raza 2023).

In 1849, the East India Company defeated the Sikh Empire and annexed the Punjab. From the mid-1880s, nine Punjab Canal Colonies were brought under cultivation. Feeding off the Indus River (and its tributaries), centralized irrigation schemes involving large-scale canals were introduced by British engineers to encourage the production and export of cotton, indigo, sugar, and wheat (Raza 2023). Arid subsistence production was replaced and the aesthetic, cultural, ecological, and spiritual values that the pastoralists ascribed to water were occluded by that self-servingly modernizing view. After the independence of Pakistan in 1947, these irrigation schemes expanded well into the 1970s as veritable monuments to engineering technology with the support of major international donors: the schemes continued to attract tenant farmers leasing land, including the erstwhile *Rodh Kohi* pastoralist tribes who, little by little, stopped maintaining their diversion structures and embankments.

From the 1980s, the hill torrents began for the most part to course through the *Pachad* unchecked, striking the irrigation schemes, and damaging irrigated lands, crops, houses, and public/private infrastructure (Qureshi et al. 2016). Veritable flyovers were built to channel the hill torrents over the irrigation schemes. Ever since, structural "solutions" to these interconnected issues have kept government officials and agencies, major international donors, consultants, and contractors in business with feasibility studies and pilot projects, raising the flood risk with river embanking and transferring it to the lower course of the Indus River (Raza 2023). To this day, more is known about the management strategies of farmers than about how the wider system might work better to assist farmers from a socio-centric perspective (Merrey 1986). Too often, instability and uncertainty rather than stability and certainty are the progeny of efforts to address environmental issues with problem-solving ability, efficiency, and managerial intellect. The centralized irrigation schemes that Pakistan inherited from colonialism—and was obliged to look after under pressure from political elites—disturbed a hitherto balanced human and physical geography to ultimately leave the country's poor more vulnerable to critical trends, seasonalities, and shocks (Raza 2023).

The demise of hill torrent irrigation in southern Punjab, Pakistan illustrates some of the nefarious effects that colonial rule had on indigenous production systems at the hands of such countries as Belgium, France, Germany (until World War I), Great Britain, Portugal, and Spain, notably—but not exclusively as we saw—as a result of the Scramble for Africa (ca. 1884–1914). (Colonialism has existed since ancient times but is most closely associated with the process of European settlement and political control that extended across Africa, the Americas, parts of Asia, and Australia until the mid-1900s.) There is more reason to compare than to contrast: in many instances, large areas were expropriated and ascribed to the production of cash crops for export, ostensibly to the benefit of the metropole. Cotton, indigo, sugar, and wheat were favored in southern Punjab, Pakistan; across the African continent, the colonizers cultivated cocoa, coffee, cotton, groundnuts, oil palm, rubber, sisal, sugar, tea, tobacco, and vine over large areas of land, upsetting ecological balances (Stock 2013). In southern Africa, for example in Kenya, Rhodesia (now Zimbabwe), and South Africa, the colonizers outdid themselves: they ousted farmers and pastoralists and replaced them with European labor controlling large farms and grazing pastures on freehold. Commonly, colonization ruptured ecosystems, not forgetting the cultures and livelihoods of indigenous populations: its consequences (e.g., climate change, deforestation, desertification, economic disparities, extermination of species, flooding, introduction of diseases, loss of language and culture, political turmoil, pollution, poverty, salinization, social inequalities, urbanization, waterlogging) are being felt to this day in the former colonies and, indeed, across the globe.

Note. The data in this précis was sourced from sundry pages on Dera Ghazi (D.G.) Khan District, Pakistan, the Punjab Canal Colonies, Rajanpur District, and *Rodh Kohi* in Brittanica (https://www.britannica.com/) and Wikipedia (https://www.wikipedia.org/).

References

Kamran M, Shamshad T (2014) Impacts of hill torrents' management on socio-economic conditions of arid land farmers: a case study of Tehsil D.G. Khan. Asian J Humanity Art Lit 1(3):155–163

Merrey D (1986) The local impact of centralized irrigation control in Pakistan. International Water Management Institute. https://ideas.repec.org/h/ags/iwmibc/208369.html

Qureshi M, Rashid M, Shah M, Baluch M (2016) Hill torrent management in southern Punjab of Pakistan: historic perspective and new trends. University of Engineering and Technology Taxila. Tech J. Taxila 21(2):2–9

Raza S (2023) Engineering disaster and the legacy of colonialism in Pakistan's floods. Red Pepper, January 18. https://www.redpepper.org.uk/Pakistan-flood-climate-change-colonialism-british-empire-imperialism-green-revolution/

Shafiq M (2012) Hill torrent management initiatives in southern part of Punjab: an overview, impact analysis, and way forward. 72nd Annual Session of Pakistan Engineering Congress, Paper No. 746, 289–314. https://docslib.org/doc/10467982/hill-torrent-management-initiatives-in-southern-part-of-punjab-an-overview-impact-analysis-way-forward

Stock R (2013) Africa south of the Sahara: a geographical interpretation, 3rd edn. Guilford Press

Zahid U (2023) An overview of the irrigation system of Pakistan, February 7. https://www.graana.com/blog/an-overview-of-the-irrigation-system-of-Pakistan/

Chapter 26
Understanding the Environment in Africa: A Start-Up Kit

Abstract Much as elsewhere, the environment in Africa is shaped by natural events and human activities. That said, the colonization of Africa inflicted profound damage on indigenous societies and the environment and set in train forms of uneven development that persist to this day. With an accent on geography, this précis aims to promote understanding of the environment in Africa.

26.1 Why It Is Important to Study the Environment in Africa and What Practical Approach Might One Adopt to That Intent?

Africa is a vast continent with a rapidly growing population of close to 1.5 billion that is projected to reach about 2.5 billion in 2050. At present, Africa's population makes up almost 18% of the world's population; in 2050, more than 25% of human beings on the Earth will be African. Many countries in sub-Saharan Africa are mired in poverty, with a rapidly declining resource base especially with respect to forest resources. Distressingly, because of its absolute dependence on climate-related activities and products, Africa is more vulnerable than any other region to the world's changing weather patterns under all climate scenarios above 1.5°C. Africa's capacity for adaptation to climate change is low: the changing environment will become an ever more significant factor contributing to food insecurity, stress on water resources, and population displacement, thereby threatening to undo the continent's modest development gains. At the same time, Africa is integral to the world economy: what happens in Africa already influences Europe, the Middle East, and Africa's trading partners, and will increasingly continue to do so. The foregoing underscores the importance of studying the environment in Africa so that one might design and locate appropriate interventions and responses. In any given region of Africa, unique environmental characteristics challenge researchers and policy makers: the most practical approach to study and understanding is that of environmental geographers who look at the cultural, spatial, and temporal aspects of the

O. Serrat, *Anthropogenic Solutions for Climate Change*,
https://doi.org/10.1007/978-981-96-8654-4_26

environment (e.g., customs, economic and social development, ethnic composition, food, language, natural resources, religion).

26.2 What Branches of Knowledge Can Help Frame Our Understanding of the Environment in Africa?

The study of the environment takes place at the intersection of human and physical geography. Human geography preoccupies itself with the spatial relationships between human communities, cultures, and economies, including their connections to the environment. Physical geography (or physiography) examines the processes and phenomena that shape and condition the Earth's surface, including the animals and plants that populate it. As might be expected, the study of the environment in Africa as elsewhere turns on many fields of study: agriculture, anthropology, biology, climatology, ecology, economics, the humanities, hydrology, politics, and sociology come to mind, but should in every instance be informed by geography. Bridging disciplines is vital to the study of the environment: considering issues from the perspective of only one branch of knowledge cannot do justice to the complexity and importance of the endeavor.

26.3 What Major Subsystems (or "Spheres") Encompass the Earth's Environment? What Roles Did They Play in the Colonization of Africa?

Four "spheres" overlap to form the Earth's environment: (a) the hydrosphere (water); (b) the biosphere (living things); (c) the lithosphere (land); and (d) the atmosphere (air). Alone and in synergy, these four "spheres" differentiate Africa from other continents by their unique quality, magnitude, and interactions. The colonization of Africa was sparked in the main by the continent's rich lithosphere (e.g., copper, diamonds, gas, gold, oil, salt, silver, wood); Africa's biosphere attracted lesser but equally predatory interest (e.g., animal hides and trophies, ivory from the African elephant, horns from the African rhinoceros). Aside from sheer greed, feelings of superiority, or the perceived need to spread Christianity and the alleged benefits of European civilization, the prime driver of the imperialist "Scramble for Africa" was industrial development in Europe, which required ever-greater resources not available there.

26.4 How Did Colonization Impact Africa?

From about 1870, Britain began to deploy systematic efforts to source raw materials from and build lucrative markets in Africa, quickly followed by the French. In 1884–1885, the Berlin Conference regulated European colonization and trade and sparked the "Scramble for Africa." By 1900, much of Africa had been colonized by seven European powers: Belgium, Britain, France, Germany, Italy, Portugal, and Spain. By 1914, only Ethiopia and Liberia remained independent. The impact of colonization on Africa and its peoples cannot be underestimated: with highly detrimental effects to this day, including extreme fragmentation of the political map, the colonization of Africa brought about the re-structuring of land ownership and utilization. In that way, Africa became the primary source of raw materials cultivated, mined, or seized by the colonialists: this led to resource depletion and labor exploitation and curtailed the continent's own industrialization. Past mining enclaves and under pressure to make their territories pay for themselves, the colonial powers subjugated Africa to the production of cocoa, coffee, cotton, groundnuts, palm, rubber, sugarcane, tea, and other profit crops. From a small local base, production, transport, and trade expanded vastly. Cash cropping engendered dual economic structures (viz., market economy vs. subsistence economy), each of which resting on quite different reward systems. Before long, cash cropping caused widespread and severe dislocation of indigenous societies: woefully, it warped traditional interactions between indigenous societies and their environment and impacted value structures, including moral attitudes toward nature and the erstwhile accent on wisdom. Colonization damaged Africa culturally, economically, and politically: its aftereffects are felt to this day and include large-scale conversion of forests to monocultures; the attribution of the best land to economic elites; ethnic or tribal rivalry as well as civil unrest from the creation of "artificial" countries that did away with chieftainships and kingdoms; new forms of dependence from coercive cultivation and mining schemes that inhibited diversification; urbanization; poor health and education; and weakened governance, to cite a few.

26.5 What Forms of Land Tenure Did the Colonization of Africa Introduce?

Land tenure has always been a central force behind the formation of cultural, economic, social, and political institutions. Prior to colonization, most land in Africa was held under customary tenure: a recognized traditional authority, viz., a chief, granted land to a person or family under customary law. With the introduction of plantations, the colonization of Africa restructured, reorganized, or reformed land tenure and land utilization to promote cash crop economies. Specifically, colonization introduced such concepts and associated practices as freehold, leasehold, and public land:

- **Freehold Land.** Freehold land grants the holder exclusive possession in perpetuity. A freeholder has the right to subdivide or lease land without seeking Government approval and the Government has no legal right to interfere with the occupational right to land. Freehold land, therefore, offers the highest security of tenure. In many countries, for example in Zimbabwe (then-Rhodesia), large plots of land were granted to foreign settlers for cultivation of cash crops and livestock production: this led to the displacement of indigenous farmers from prime to lower-quality lands.
- **Leasehold Tenure.** Leasehold tenure grants a person exclusive right to occupy and use land for a fixed period. Under both common law and civil law, such land may be leased or rented by to another party. The arrangements typically rest on very short terms; however, these can stretch to the 99-year leases that are common in the United Kingdom, allowing various degrees of freedom in the use of the property.
- **Public Land.** Public land, the third category of land ownership introduced by the colonizers, comprises land that is held in trust by Government, local, or traditional authorities and is used openly or made accessible to the public. Public land includes land that is gazetted for historical and cultural areas, national parks, recreation areas, as well as urbanization.

Post-independence, the systems of land tenure introduced by the colonial powers did not change. What with the sweeping disruptions that colonization brought about with respect to African societies and economies, it was difficult to revert in full to earlier systems. For sure, perhaps as much as 90% of land in Africa is still held under customary tenure: but, this means it continues to lack legal recognition, thereby discouraging investment to the detriment of agricultural productivity; and so, with rapid demographic growth, the livelihoods of entire populations are at increasing risk. Moreover, the most favorable lands are concentrated into large holdings that are controlled by national governments or economic elites: this means that poor rural farmers conduct subsistence farming on small holdings that have marginal soils.

26.6 Why Is Agricultural Productivity in Africa Low?

Africa has tropical or equatorial climates, dry or arid climates, moderate or temperate climates, and highland climates. The two major types of climate in Africa are tropical or equatorial climates and dry or arid climates but most of Africa lies within the tropical region, which is bounded in the north by the Tropic of Cancer (approximately 23.2 degrees north of the terrestrial Equator) and bounded in the south by the Tropic of Capricorn (approximately 23.2 degrees south of the terrestrial Equator). The tropical or equatorial climate is characterized by humidity, warm temperatures, and high precipitation (viz., rain). Some regions have a dry–wet season; other regions receive consistent rainfall all year-round. The drawback of agriculture in the tropics is that soils are inherently of low fertility because nutrients are continually leached by heavy rainfall, which puts a cap on fertilizer use efficiency.

26.7 How Do Definitions of Natural Resources Condition Their Treatment?

How one defines natural resources depends on what discipline (or school of thought) one subscribes to, the predominant theories being economics (economic theory) on the one hand and environmental science (ecosystem theory) on the other. Economic theory treats natural resources as material sources of wealth. In contrast, ecosystem theory deems that nature has actual (or potential) ecological value to life: therefore, it integrates a range of biological, existential, moral or ethical, pragmatic, and recreational concerns. Sped by the quest for profit, management strategies for natural resources have for long been submitted to economic theory. But, the world now faces unprecedented and interconnected environmental challenges in areas including biodiversity loss, climate change, deforestation, food and water insecurity, melting ice caps and sea level rise, ocean health, overfishing, pollution, and soil degradation. With the adoption by the United Nations General Assembly in 2015 of the 2030 Agenda for Sustainable Development and its 17 Sustainable Development Goals, ecosystem theory is belatedly attracting attention with pledges to improve and protect the quality of the natural environment through changes to environmentally harmful anthropogenic activities; adoption of economic, political, and social organization that conduces the benign treatment of the environment; and long-overdue reassessment of humankind's relationship with nature. The 2030 Agenda is a plan of action for people, planet, and prosperity. Today, ecosystem theory is an integrated and interdisciplinary approach with which to put people and their natural resource use practices squarely at the center of decision-making.

Historically, Africans never considered land as a commodity to be traded. African traditions were—and remain—fundamentally anchored in subsistence farming, with limited interest in market forces and the values that capitalism advocates and thrives on (i.e., competition, freedom to choose, limited role of government, market mechanism, private property, self-interest). Specifically, in Africa, land has for long been treated as communal, not individual. But, there is more: Africans also consider that land has cultural, ecological, social, and spiritual values; that it belongs equally to the living, the dead, and the unborn; and that it provides a living connection to ancestors whose home it is. In Africa, therefore, the protection of natural resources, the environment, and wildlife has historically been based on moral or ethical aspects; it follows that ecosystem theory is more closely aligned to what the communities in that region need.

26.8 What Would a Deterministic View of the Relationship Between Poverty in Africa and Natural Resources Management Imply? What Are the Limits of Determinism?

Determinism would have it that all is pre-ordained by way of causal relationships, meaning, that everything flows from some action or set of actions taken upstream in time. In relation to poverty, a deterministic perspective would advance that—in a vicious downward spiral—poverty is the prime cause and effect of problems. From that perspective, for instance, natural resources are being actively degraded by the poor in Africa: thus, determinism would have it that the poor overexploit natural resources to survive, which leads to further degradation. Determinism can only reduce understanding and is ultimately fatalistic. Deterministic perspectives do not make for knowledge creation and knowledge transfer, something that is fundamental to learning, scholarly enterprise, and science; debates are better framed by the hard study of actors, context, region, situation, time, and such like factors or drivers. Ideally, scholars and practitioners should interpret situations based on observations, inductions, and inferences, and refrain from shortsightedly explaining circumstances with prefabricated—and often imported—models.

26.9 What Key Distinguishable Attributes of Law, Policy, and Governance Condition Natural Resource Management?

In Africa as elsewhere, law, policy, and governance are eminently relevant to the management of natural resources. The matter of law is relatively straightforward: law embodies legislation that compels or prohibits certain actions or behaviors by individuals and groups in society. Next, policy is a deliberate plan of action to guide decisions and achieve rational outcomes, be they local, national, international, regional, or even global; it follows that policy can have spatial and temporal dimensions. Given the importance of communal rights in Africa, there is much reason and scope to improve natural resource management through policy objectives, policy instruments, and policy rules; respectively, these have to do with strategy, operations, and tactics. But, governance is more complex: to wit, the act or process of governance refers to the exercise of political authority and the legal use of government instruments (such as institutions and resources) to manage affairs. It follows that better management of natural resources in Africa as elsewhere must be based on what judicious mix of law, policy, and governance is called for: there are deep implications for democratic processes, institutions, and participation.

26.10 How Might Policy Underpin Better Natural Resources Management?

The ingredients of policy for natural resources management are (a) policy objectives that help steer decision-making at all levels of government; help avoid conflict, misinterpretation, misunderstanding, and uncertainty; and facilitate planning and budgeting; (b) policy instruments that through effective institutions enable government interventions to accomplish goals or solve problems through the exercise of power and authority; and (c) policy rules that define the exact means of operation (viz., who can use what, where, when, and how). Policy objectives, policy instruments, and policy rules must be considered and framed as a coherent whole, else policy will be ineffective. Certainly, for instance, focusing on policy rules will be counterproductive if policy objectives are poorly formulated or policy instruments work at cross-purposes. In short, the ingredients of policy must work together.

2.16 How Might Policy Underpin Better Natural Resources Management?

[illegible]

Correction to: Anthropogenic Solutions for Climate Change

Correction to: Chapters 22 and 25 in: O. Serrat, *Anthropogenic Solutions for Climate Change*, https://doi.org/10.1007/978-981-96-8654-4

The original version of the book was inadvertently published with the below errors. The corrections have been carried out in the chapters.

Chapter 22: Preserving Paradise: The Heart of Borneo Initiative

On page 223, the text "defense spending" has been corrected to "deforestation" in lines 6, 10, 13 and 40.

On page 224, the text "defense spending" has been corrected to "deforestation" in line 15.

Chapter 25: The Demise of Hill Torrent Irrigation in Southern Punjab, Pakistan

On page 259, the line "With about 243 million people (2023), however, Pakistan is home to the world's fifth biggest population (and its biggest Muslim population). "has been corrected to" With about 243 million people (2023), however, Pakistan is home to the world's fifth biggest population (and its biggest Muslim population)."

The updated version of these chapters can be found at
https://doi.org/10.1007/978-981-96-8654-4_22
https://doi.org/10.1007/978-981-96-8654-4_25

O. Serrat, *Anthropogenic Solutions for Climate Change*,
https://doi.org/10.1007/978-981-96-8654-4_27

Index

O. Serrat, *Anthropogenic Solutions for Climate Change*,
https://doi.org/10.1007/978-981-96-8654-4

MIX
Papier aus verantwortungsvollen Quellen
Paper from responsible sources
FSC® C105338

If you have any concerns about our products,
you can contact us on
ProductSafety@springernature.com

In case Publisher is established outside the EU,
the EU authorized representative is:
Springer Nature Customer Service Center GmbH
Europaplatz 3, 69115 Heidelberg, Germany

Printed by Libri Plureos GmbH
in Hamburg, Germany